Lecture Notes in Computer Sc

T0250559

Commenced Publication in 1973
Founding and Former Series Editors:
Gerhard Goos, Juris Hartmanis, and Jan van Leeuwen

Lecture Notes in Computer Science

Herbert Leitold   Evangelos Markatos (Eds.)

Communications and Multimedia Security

10th IFIP TC-6 TC-11 International Conference, CMS 2006
Heraklion, Crete, Greece, October 19-21, 2006
Proceedings

 Springer

Volume Editors

Herbert Leitold
Secure Information Technology Center - Austria
Inffeldgasse 16a, 8010 Graz, Austria
E-mail: Herbert.Leitold@a-sit.at

Evangelos Markatos
Foundation for Research and Technology - Hellas
Institute of Computer Science
Heraklion, Crete, 71110 Greece
E-mail: markatos@ics.forth.gr

Library of Congress Control Number: 2006934468

CR Subject Classification (1998): C.2, E.3, F.2, H.4, D.4.6, K.6.5

LNCS Sublibrary: SL 4 – Security and Cryptology

ISSN   0302-9743
ISBN-10   3-540-47820-5 Springer Berlin Heidelberg New York
ISBN-13   978-3-540-47820-1 Springer Berlin Heidelberg New York

Springer is a part of Springer Science+Business Media

springer.com

© IFIP International Federation for Information Processing 2006
Printed in Germany

Typesetting: Camera-ready by author, data conversion by Scientific Publishing Services, Chennai, India
Printed on acid-free paper   SPIN: 11909033   06/3142   5 4 3 2 1 0

Preface

During the last few years we see network and information system security playing an increasingly important role in our everyday lives. As our computers continue to get infested by all sorts of malware, and as our networks continue to choke with spam and malicious traffic, we see more and more people losing their confidence in information technologies as they get significantly concerned about their security as well as their privacy and that of their loved ones. In their effort to cope with the problem, scientists, managers, and politicians all over the world have designed and are currently implementing systematic approaches to network and information security, most of which are underlined by the same principle: *there is much more room for improvement and research.*

Along the lines of encouraging and catalyzing research in the area of communications and multimedia security, it is our great pleasure to present the proceedings of the 10th IFIP TC-6 TC-11 Conference on Communications and Multimedia Security (CMS 2006), which was held in Heraklion, Crete on October 19-21, 2006. Continuing the tradition of previous CMS conferences, we sought a balanced program containing presentations on various aspects of secure communication and multimedia systems. Special emphasis was laid on papers with direct practical relevance for the construction of secure communication systems. The selection of the program was a challenging task. In total, we received 76 submissions, from which 22 were selected for presentation as full papers.

We want to thank all contributors to CMS 2006. In particular, we are grateful to the authors and invited speakers for contributing their latest work to this conference, as well as to the PC members and external reviewers for their critical reviews of all submissions. Finally, special thanks go to the Organizing Committee who handled all local organizational issues and provided us with a comfortable location and a terrific social program. For us, it was a distinct pleasure to serve as Program Chairs of CMS 2006. We hope that you will enjoy reading these proceedings and that they will be a catalyst for your future research in the area of communications and multimedia security.

October 2006   Herbert Leitold and Evangelos Markatos
Program Co-chairs
CMS 2006

Organization

CMS 2006 was organized by A-SIT (Zentrum für sichere Informationtechnologie - Austria), FORTH (Foundation for Research and Technology - Hellas), and ENISA (European Network and Information Security Agency).

Program Chairs

Program Co-chair   Herbert Leitold, A-SIT, Austria
Program Co-chair   Evangelos Markatos, FORTH-ICS, Greece
Organizing Chair   Angelos Bilas, FORTH-ICS, Greece

Program Committee

Andre Adelsbach, Horst Görtz Institute for IT Security, Germany
Elisa Bertino, Purdue University, USA
Carlo Blundo, Università di Salerno, Italy
Herbert Bos, Vrije Universitei, The Netherlands
David Chadwick, University of Kent, UK
Howard Chivers, Cranfield University, UK
Bart de Decker, KU Leuven, Belgium
Gwenaël Doërr, University College London, UK
Stephen Farrell, Trinity College Dublin, Ireland
Miroslav Goljan, SUNY Binghamton, USA
Dimitris Gritzalis, Athens University of Economics and Business, Greece
Patrick Horster, University Klagenfurt, Austria
Russ Housley, Vigil Security, USA
Borka Jerman-Blažič, Institut Jožef Stefan, Slovenia
Stefan Katzenbeisser, Philips Research, The Netherlands
Stephen Kent, BBN Technologies, USA
Klaus Keus, BSI, Germany
Antonio Lioy, Politecnico di Torino, Italy
Peter Lipp, Graz University of Technology, Austria
Michael Locasto, Columbia University, USA
Javier Lopez, Univeristy of Malaga, Spain
Chris Mitchell, Royal Holloway, University of London, UK
Sead Muftic, The Royal Institute of Technology, Sweden
Jose Nazario, Arbor Networks, USA
Fernando Perez-Gonzalez, University of Vigo, Spain
Günther Pernul, University of Regensburg, Germany
Reinhard Posch, Graz University of Technology, Austria
Bart Preneel, KU Leuven, Belgium

Wolfgang Schneider, Fraunhofer Institut SIT, Germany
Frank Siebenlist, Argonne National Laboratory, USA
Otto Spaniol, Aachen University of Technology, Germany
Leon Strous, De Nederlandsche Bank, The Netherlands
Panagiotis Trimintzios, ENISA, EU
Vincent Rijmen, Graz University of Technology, Austria
Andreas Uhl, University of Salzburg, Austria
Von Welch, National Center for Supercomputing Applications, USA
Vinod Yegneswaran, University of Wisconsin, USA
Claus Vielhauer, OttovonGuericke University Magdeburg, Germany

Local Organization

Angelos Bilas
Yiannis Askoxylakis
Theodossia Bitzou
Eleni Orphanoudakis

External Referees

Isaac Agudo
Lejla Batina
Abhilasha Bhargav-Spantzel
Ji-Won Byun
Martin Centner
Emanuele Cesena
Peter Danner
Liesje Demuynck
Wolfgang Dobmeier
Michail Foukarakis
Theo Garefalakis
Steven Gevers
Sotiris Ioanidis
Mario Ivkovic
Jongsung Kim
Tobias Kölsch
Jan Kolter
Franz Kollmann
Stefan Kraxberger
Mario Lamberger
Ioannis Marias
Jose A. Montenegro
Vincent Naessens
O. Otenko

Andriy Panchenko
Udo Payer
Lexi Pimenidis
Dimitris Plexousakis
Stefan Rass
Thomas Rössler
Christian Schläger
Martin Schaffer
Peter Schartner
Rolf Schillinger
Stefaan Seys
Mohamed Shehab
Adam Slagell
Arne Tauber
Peter Teufl
Tine Verhanneman
Kristof Verslype
Ivan Visconti
Ralf Wienzek
Li Weng
Yongdong Wu
Hongjun Wu
Kostantinos Xinidis

Sponsoring Institutions

A-SIT (Zentrum für sichere Informationtechnologie - Austria)
FORTH (Foundation for Research and Technology - Hellas)
ENISA (European Network and Information Security Agency).

Table of Contents

Advances in Network and Information Security

Computing of Trust in Ad-Hoc Networks

Huafei Zhu, Feng Bao, and Jianwei Liu*

Institute for Infocomm Research, A-star, Singapore
{huafei, baofeng}@i2r.a-star.edu.sg, liujianwei@buaa.edu.cn

Abstract. Although, the notion of trust has been considered as a primitive for establishing relationships among nodes in ad-hoc networks, syntax and metrics of trust are not well defined. This paper studies computing of trust in ad-hoc networks and makes the following three contributions. Firstly, the notion of trust is formalized in terms of predict functions and strategy functions. Namely, the notion of trust in this paper is defined as a predict function that can be further evaluated by a strategy function for a pre-described action; Secondly, structures of trust are formalized as a map between a path in the underlying network graph and the corresponding edge of its transitive closure graph; Thirdly, a generic model for computing of trust in the small world is proposed.

Keywords: Ad-hoc network, Transitive graph, Trust computing.

1   Introduction

Ad-hoc networks formed by a set of dynamic nodes without relying on a preexisting infrastructure have been a very attractive field of academic and industrial research in recent years due to their potential applications and the proliferation of mobile devices. For example, a set of self-organized nodes are selected to accomplish a designated task say, collaboratively computing a multi-variable boolean function $f(x)$ on input x. In this setting, all nodes involved in the computation of $f(x)$ have to access a certain resource to obtain data in order to complete the task. As a result, a node should prove its membership to a self-organized set which is supposed to have access to the resource. If traditional public key infrastructures (PKI) are assumed, then the authentication of membership should be an easy task. However, it is difficult to deploy centralized certification authorities in ad-hoc networks due to the lack of central services.

Trust is considering a primitive for the establishment of relationship in ad-hoc networks. In our opinion, Alice trusts Bob means that Alice predicates that Bob will act on some action honestly in the future. It follows that the notion of trust should be defined as a predict (by $\mathcal{PT}$, we denote the function of a prediction). For example, a verification of a signature is a predict function; If an output of the predict function is 1, Alice's trust value evaluation strategy (by $\mathcal{SG}$, we denote a strategy for evaluating trust value) is then performed. The output value is

---

* BeiHang University, China.

H. Leitold and E. Markatos (Eds.): CMS 2006, LNCS 4237, pp. 1–11, 2006.

called trust degree (or trust value) of Alice to Bob for the pre-specified action. Intuitively, the output of $\mathcal{SG}$ satisfies the following properties:

- one-wayness: for a fixed action $\mathcal{A}$ (by $\mathcal{A}$, we denote an action chosen from the pre-described action space which is denoted by $\mathcal{A}^*$), the concept of trust is one-way (or asymmetric) in the sense that N_1 trusts N_2's action $\mathcal{A}$ does not imply that N_2 trusts N_1's action $\mathcal{A}$.
- transitivity: the concept of trust maintains transitivity for a fixed action. That is, if S trusts N_1's action $\mathcal{A}$, and N_1 trusts N_2's action $\mathcal{A}$, and N_2 trusts T's action $\mathcal{A}$, then S trusts T's action $\mathcal{A}$. We stress that the action $\mathcal{A}$ specified by the source node S, intermediate nodes and the target node T must be same, otherwise there is no reason to maintain the transitivity.

If we view individual participant in ad-hoc networks as a node of a delegation graph G, then a mapping between a delegation path from the source node S to the target node T in the graph G and an edge in the transitive closure graph G^* of G can be established. We thus study the following fundamental research problems: how to evaluate trustworthiness of participants in an edge of G? how to compute trustworthiness of a recommendation path of G? Namely, how to evaluate the trustworthiness of edges in the transitive closure graph G^* of G?

1.1 Previous Works

The pioneer work for computing of trust is due to Beth, Borcherding and Klein [2] and Yahalom, Klein and Beth [12]. In their seminal papers, models for computing of trust in distributed network are outlined. Although, a collection of genuine ideas were presented, there was no formal definition of trust presented in their papers. Following their seminal contributions, Zhu et al [14] distilled transitivity of trust by means of transitive graph and then applied their results for computing of trust in wireless networks (e.g., [15], [16] and [7]). Although, these are interesting applications of trust in the real world, the notion of trust is not well defined. For example, the term trust (and trust value/degree) defined in their previous papers does not cover the following important issues: the formalization of the notion of action (and action space), and the notion of trust; and the longer size of a recommendation path, the less trust value along a path; We stress that these issues are inherent properties of the notion of trust, and thus must be satisfied. As a result, any more satisfactory solution for computing of trust is certainly welcome.

1.2 This Work

The contributions of the paper are three-fold. In the first fold, the notion of trust is formalized in terms of predict functions and strategy functions. Namely, the notion of trust is defined as a predict that can be further evaluated by a strategy function for a pre-described action if a predict outputs 1; In the second fold, the structures of trust is formalized as a mapping between a path in a network graph G and an edge of the transitive closure graph G^* of G. In the third fold, a generic model for computing of trust in the small world phenomena is proposed.

The remainder work of this paper is organized as follows: In Section 2, syntax, structure of trust are introduced and formalized. In Section 3, a framework for computing of trust in ad-hoc works is proposed and analyzed. We propose an example for computing of trust in the small world phenomena in Section 4, and conclude our work in Section 5.

2   Trust: Syntax, Characteristics and Structures

2.1   Definition of Trust

Tons of definitions regarding trust have been presented in the literature. The commonly cited definition of trust is due to Golbeck[4]: Alice trusts Bob if she commits to an action $\mathcal{A}$ based on a belief that Bob's future actions will lead to a good outcome. We stress that Golbeck's definition does not capture the prediction of trust. That is, the notion of trust should be defined binary values: trust (a predict $\mathcal{PT}$ outputs 1) or distrust (a predict $\mathcal{PT}$ outputs 0). In case of trust (or distrust), we can talk about the degree of trust (or distrust). Since the notion of trust and the notion of distrust are complementary concepts, it is enough for us to define the concept of trust.

We also stress that an action $\mathcal{A}$ should be sampled by any probabilistic polynomial time (PPT) Turing machine on input of a system parameter k. That is, on input of a system parameter k, the PPT Turing machine will specify an action space ($\mathcal{A}^*$) such that on input of an index $i \in I$, an action $\mathcal{A}_i \in \mathcal{A}^*$ is selected.

Given an action $\mathcal{A} \in \mathcal{A}^*$, Alice runs a predict function $\mathcal{PT}$ which outputs 0 or 1. Once $\mathcal{PT}(\mathcal{A})=1$, Alice can preform her strategy function $\mathcal{SG}$ to obtain a trust value with the help of her auxiliary information aux (intuitively, the auxiliary information aux is a cumulative history record of Bob maintained by Alice herself).

Thus, to formalize the notion of trust, we first need to provide a formal definition of an action. Let $\mathcal{A}$ be a disjunction $c_1 \vee \cdots \vee c_m$ of clauses, where each clause c_i is a conjunction $l_1 \wedge \cdots \wedge l_{t_i}$ of t_i literals. Each literal l_j is either a Boolean variable X_i or its negation $\bar{X}_i$. Without loss of generality, we may assume that each variable occurs at once in any given clause.

Definition 1. *An action $\mathcal{A}$ is a disjunctive normal form over k Boolean variables $X_1, \cdots, X_k$. The set of all actions is call action space which is denoted by $\mathcal{A}^*$.*

To define the trust value of an action, we need to make the following assumptions:

- the underlying network is an unknown fixed-identity graph G, where each node has a unique identity N_i which cannot be forged. And each node knows the identities of its neighbors in G. Such an assumption is necessary since if a node forges its node id, then it is impossible for one to distinguish a forged id from a genuine id (as there is no public key infrastructure assumption involved in our model);

- a keyed-identity of node N_i is of form $k_i := (N_i, g(N_i))$ where $g(N_i)$ is a claimed public key of the node N_i.

Definition 2. *Let k_A and k_B (for convenience, we sometime will write k_A simply as A) be two nodes in a graph G. An auxiliary information $aux^A(B) \in \{0,1\}^{poly(\lambda)}$ is a string that cumulatively records the state of B by A.*

Definition 3. *An auxiliary information is called samplable if there is a deterministic polynomial time algorithm $\mathcal{I}$ such that on input λ, k_A and k_B, it outputs $aux^A(B) \in \{0,1\}^{poly(\lambda)}$. By $\mathcal{I}^*$, we denote operators set $\mathcal{I}$.*

Definition 4. *On input k_A and k_B, an action $\mathcal{A} \in \mathcal{A}^*$, and auxiliary information $aux^A(B)$, a deterministic predict function $\mathcal{PT}$ outputs a bit $b \in \{0,1\}$. Once $\mathcal{PT}$ outputs 1, $\mathcal{PT}$ then runs a trust evaluation strategy algorithm $\mathcal{SG}$ which outputs a positive value $\alpha \in \{0,1\}$. This value α is called a trust value of k_A regarding the action $\mathcal{A}$ associated with k_B.*

2.2 Trust Structures

Definition 5. *A graph $G = (V,E)$ has a finite set V of vertices and a finite set $E \subseteq V \times V$ of edges. The transitive closure $G^* = (V^*, E^*)$ of a graph $G = (V,E)$ is defined to have $V^* = V$ and to have an edge (u,v) in E^* if and only if there is a path from u to v in G.*

Based on the above assumptions, we can now define the structure of trust. For a given path $S \to N_1 \to \cdots \to N_k \to T$, we define the trust values of individual edges $S \to N_1$, $N_1 \to N_2$, $\cdots$ and $N_k \to T$. And we then compute the edge $S \to T$ in the transitive closure graph G^*. A a result, two types of trust structures can be defined: a direct trust and a recommended trust. Intuitively, a direct trust is an edge between two nodes in a graph G while recommended trust is an edge defined in its corresponding transitive closure graph G^*. As a result, the notion of recommended trust can be viewed as a natural extension of the notion of the direct trust (if the number of intermediate nodes in a path is zero). Generally, for any path of length k defined over G, a recommended trust RT is defined of the following form: $\Pi_{i=1}^{k} DT_i$, where DT_i is a direct trust of N_i to N_{i+1}.

3 Computing of Trust

3.1 Computing of Direct Trust Values

Let $dtv^A(B)$ be a direct trust value assigned to B by A; The range of $dtv^A(B)$ is $[0,1]$. If the trust value $dtv^A(B)$ is 0, it means that A does not trust B at all; if $dtv^A(B)=1$, it means that A trusts B completely; if $dtv^A(B) = \alpha$, it means that A trusts B with degree α, where $\alpha \in (0,1)$. Computing of direct trust value $dtv^A(B)$ can be performed as follows:

- Input $\Theta := (k_A, k_B, aux^A(B), \mathcal{A})$, where k_A (resp. k_B) is a key-identity of node A (resp. B) and $aux^A(B)$ is auxiliary information regarding the node B maintained by the node A;

- Computing $u \leftarrow \mathcal{PT}(\Theta)$;
  - if u=0, then $v \leftarrow 0$;
  - if u=1, then $v \leftarrow \mathcal{SG}(\Theta | u = 1)$
- Output $dtv^A(B) \leftarrow v$.

We stress that the above computation of the direct trust value captures two things. The first one is the notion of predict. This means that A either trusts B or distrusts B. The second one is the computation of direct trust value under the condition that A trusts B.

3.2   Computing of Recommended Trust Values over Bounded-Disjoint-Paths

Suppose $p_1, \cdots, p_k$ be k paths connected between S and T. These paths are referred to as delegation paths. Let $N^i = \{N_1^i, \cdots, N_{l_i}^i\}$ be a set of intermediate recommenders (not including S and T) in the path p_i.

Definition 6. *Two paths from S to T, say $S \to N_1^i \to \cdots \to N_{l_i}^i \to T$ and $S \to N_1^j \to \cdots \to N_{l_j}^j \to T$ are disjoint if $N_a^i \neq N_b^j$, for all a, $1 \le a \le l_i$ and all b, $1 \le b \le l_j$.*

Definition 7. *Suppose $p_1, \cdots, p_k$ be k paths connected between S and T, $p_1, \cdots, p_k$ are called mutually disjoint if paths are pair-wise disjoint.*

Definition 8. *A path p is ρ-bounded if its length is at most ρ.*

Given a directed graph G (we distinguish the node S and the node T) and a path bound ρ, we are interested in finding the maximum set of mutually disjoint ρ-bounded paths from S to T − an interesting research problem first introduced and formalized by Reiter and Stubblebine in [9], where the Bounded-Disjoint-Paths (BDP) problem is shown to be difficult if P ≠ NP. As a result, there is no polynomial approximation algorithm APP for BDP such that $\text{BDP}((G, \rho, S, T)$ -$\text{APP}(G, \rho, S, T) \le C$ for a fixed constant C. This means that it is hard for one to find almost bounded disjoint paths in the graph G. Thus, for computing of trust in ad-hoc networks, we only consider a set of incomplete Bounded-Disjoint-Paths. As a result, to define the trust value for a set of bounded disjoint paths (say, $p_1, \cdots, p_k$), we need to consider the following two cases:

- Case 1: given a path p=$\{N_1, \cdots, N_l\}$ (excluding the source node S and the target node T), how to define the trust value associated with the path p?
- Case 2: given a collection of paths(say, $p_1, \cdots, p_k$), how to define the trust value associated with the paths?

To compute trust value in Case 1, we first informally define the recommended trust value of S to T by the following formula for a given path p=: $\{S, N_1, \cdots, N_l, T\}$:

$$rtv^S(T, p) = dtv^S(N_1) \diamond dtv^{N_1}(N_2) \diamond \cdots \diamond dtv^{N_{l-1}}(N_l) \diamond dtv^{N_l}(T)$$

We stress that the direct trust value $dtv^{N_{i-1}}(N_i)$ has been defined in the last section. The remaining question is thus to define the exact meaning of the operator $\diamond$. Intuitively, a larger size l implies that the smaller recommended trust values $rtv^S(T)$. Furthermore, if there is a faulty node that provides a fault recommendation, the resulting recommended trust value should be low. Consequently, the operator $\diamond$ can be defined in a simple way: $x \diamond y = min\{x, y\}$.

To compute the trust value in Case 2, we first introduce the following notations. By $min\{a_{i,1}, a_{i,2}, \cdots, a_{i,l_i}\}$, we denote the recommended trust value of p_i, i.e. $rtv^X(Y, p_i) = min\{a_{i,1}, a_{i,2}, \cdots, a_{i,l_i}\}$. By $max_{i=1}^t rtv^X(Y, p_i)$, we denote the recommended trust value computed from the path set $\{p_1, \cdots, p_t\}$. The recommended trust value computed from $\{p_1, \cdots, p_t\}$ is defined below

$$rtv^X(Y, p_1, \cdots, p_t) = max_{i=1}^t min_{j=1}^{l_i}\{a_{i,j}\}$$

We stress that the recommended trust value defined above captures the intuition of the trust value:

- if there is $dtv^{N_{i-1}}(N_i) = 0$, then $rtv^S(T) = 0$; This means that if there is a fault node in a given path, the recommendation path should not be trusted at all.
- if $p' = p \cup \{N_{k+1}\}$, then $rtv^S(T, p') \leq rtv^S(T, p)$, where $p = \{N_1, \cdots, N_k\}$; This means that the longer the size of a recommendation path, the less trust value should be computed from individual recommenders along the path;
- if $rtv^S(T, p)$ is a positive and $dtv^{N_k}(N_{k+1})$ is positive, then $rtv^S(T, p')$ is positive, where $p = \{N_1, \cdots, N_k\}$ and $p' = p \cup \{N_{k+1}\}$; The means that the definition of the trust value of recommendation is transitive.

3.3   Minmax Principle for Trust Metrics

We will show that the principle for computing of trust proposed above satisfies Yao's Minimax theorem[11]. As a result, the expected running time of the optimal deterministic algorithm for an arbitrary chosen input distribution is a lower bound on the expected running time of the optimal randomized algorithm for trust evaluation. This is the most significant feature of our metrics.

Let Π be a problem with a finite set Θ of input instances of fixed size (k_A, k_B, $aux^A(B)$ $\mathcal{A}$), and a finite set of deterministic algorithms $\Gamma = (\mathcal{PT}, \mathcal{SG})$. For an input $inp \in \Theta$, and algorithm $alg \in \Gamma$, let $\mathcal{T}(\Theta, alg)$ be the running time of an algorithm alg on an input inp. For probability distribution ι over Θ, and τ over Γ. Let inp_ι denote a random input chosen according to ι and alg_τ denote a random algorithm chosen according to τ. Then by Yao's Minimax theorem[11], we have the following statement

$$min_{alg \in \Gamma} E[T(inp_\iota, alg)] \leq max_{inp \in \Theta} E[T(inp, alg_\tau)]$$

In other words, the expected running time of the optimal deterministic algorithm for an arbitrary chosen input distribution ι is a lower bound on the expected running time of the optimal randomized algorithm for τ.

Remarks. We remark that in case of two paths with the same trust value, say $0.9 \diamond 0.9 \diamond 0.3 = 0.4 \diamond 0.3 \diamond 0.3 = 0.3$, we will simply compute the mean of direct trust values in the path and then choose the path with the highest value (if the values are still same for different paths, then we can choose path according to the history record of nodes in the path). We stress that an alternative to avoid this problem is to use the product operator that is restricted to the interval [0,1] (see [1] and [6] for more details). Although the product operator has all required properties claimed above, we do not know whether the product operator satisfies Yao's Minimax theorem[11] or not. This leaves an interesting research problem.

4   Computing of Trust in the Small World

The concept of small world in the context of wireless networks first studied by Helmy [5] enables a path-finder to search paths originated from a source node to a designated target node in wireless networks efficiently. Based on this observation, we provide a practical approach to compute trust in wireless networks by viewing individual mobile device as a node of a delegation graph G and mapping a delegation path from the source node S to the target node T into an edge in the correspondent transitive closure of the graph G, from which a trust value is computed.

4.1   Path-Finder

Since wireless networks typically can be formalized as a small world [5], we thus use the technique presented in [15] for our path-finder. That is, we run an initiator of a route discovery process to generate a route request, which contains the identifiers of the initiator and the target, and a randomly generated query identifier. Each intermediate node that receives the request for the first time appends its identifier to the route accumulated so far, and re-broadcasts the request. When the request arrives to the target, it generates a route reply. The route reply contains the identifiers of the initiator and the target, the accumulated route obtained from the request, and a digital signature of the target on these elements. The reply is sent back to the initiator on the reverse route found in the request. Each intermediate node that receives the reply verifies that its identifier is in the route carried by the reply, and that the preceding and following identifiers on the route belong to neighboring nodes. If these verifications fail, then the reply is dropped. Otherwise, it is signed by the intermediate node, and passed to the next node on the route (towards the initiator). When the initiator receives the route reply, it verifies if the first identifier in the route carried by the reply belongs to a neighbor. If so, then it verifies all the signatures in the reply. If all these verifications are successful, then the initiator accepts the route.

4.2   Transitive Graph and Transitive Signature in PKI Setting

Notion. Given an undirected graph G, two vertices u and v are called connected if there exists a path from u to v; Otherwise they are called disconnected. The graph G is called connected graph if every pair of vertices in the graph is connected. A vertex cut for two vertices u and v is a set of vertices whose removal from the graph disconnects u and v. A vertex cut for the whole graph is a set of vertices whose removal renders the graph disconnected. The vertex connectivity $k(G)$ for a graph G is the size of minimum vertex cut. A graph is called k vertex connected if its vertex connectivity is k or greater.

Syntax of Transitive Signatures. A probabilistic polynomial time undirected transitive signature scheme TS is specified by four polynomial-time algorithms TKG, $TSig$, $TVer$ and $Comp$ [13]:

- The randomized key generation algorithm TKG takes input 1^k, where $k \in N$ is the security parameter, and returns a pair (tpk, tsk) consisting of public key and security key of a transitive signature scheme.
- The signing algorithm $TSig$ consists of a pair of separate algorithms: a vertex/node signing algorithm $VSig$ and a edge signing algorithm $ESig$. $VSig$ is a stateful or randomized algorithm that takes input of the security key tsk and a node v_i and returns a value called certificate of node v_i which is denoted by $Cert_{v_i}$. $ESig$ is a deterministic algorithm that takes input of the security key tsk and two different nodes $v_i, v_j \in V$, and returns a value called certificate of edge $\{v_i, v_j\}$ relative to tsk. $TSig$ maintains states which it updates upon each invocation.
- The deterministic verification algorithm TVf consists of a pair of separate algorithms $(VVer, EVer)$. $VVer$ is the deterministic vertex/node certificate verification algorithm that takes input of tpk and a certificate $Cert_{v_i}$ of vertex v_i, returns either 1 or 0. $EVer$ is the deterministic algorithm that takes input of tpk and two nodes $v_i, v_j \in V$, and a certificate σ of edge $\{v_i, v_j\}$, returns either 1 or 0 (in the former case we say that σ is a valid signature of edge $\{v_i, v_j\}$ relative to tpk).
- The deterministic composition algorithm $Comp$ takes input of tpk and nodes $v_i, v_j, v_k \in V$ and values σ_1, σ_2 to return either a value of σ or a symbol $null$ indicate failure.

The Definition of Security. Associated to transitive signature scheme $(TKG, TSig, TVer, Comp)$, adversary Adv and security parameter $k \in N$, is an experiment which is denoted by $Exp_{TS,Adv}^{tu-cma}(k)$ that returns 1 if and only if Adv is successful in its attack. The experiment begins by running TKG on input 1^k to get keys (tpk, tsk). It then runs Adv, and providing this adversary with input tpk and oracles access to the functions $ESig(tsk, \cdot)$ and $VSig(tsk, \cdot)$. The oracles are assumed to maintain state or toss coins as needed. Eventually, Adv will output $(v_{i'}, v_{j'}) \in V \times V$ and some value τ'. Let E be the set of all edges $\{v_a, v_b\}$ such that Adv made oracle queries v_a, v_b, and let V be the set of all nodes v_a such that v_a is adjacent to some edge in E. We say that Adv wins if τ' is a valid signature of $\{v_{i'}, v_{j'}\}$ relative to tpk but the edge is not $\{v_{i'}, v_{j'}\}$ in the transitive closure G of a graph $G = (V, E)$. The experiment returns 1 if Adv wins and 0 otherwise. The advantage of adversary in its attack on TS is the function $Adv_{TS,Adv}^{tu-cma}(\cdot)$ defined for k by

$$Adv_{TS,Adv}^{tu-cma}(k) = \Pr[Exp_{TS,Adv}^{tu-cma}(k) = 1]$$

We say that a transitive signature scheme is transitively unforgeable under adaptive chosen-message if $Adv_{TS,Adv}^{tu-cma}(k)$ is negligible for any adversary Adv whose running time is polynomial in the security parameters k.

Known Implementation in the PKI Setting. There is an efficient implementation of transitive signature presented in [13], which is sketched below:

- System parameters: Let p, q be two large safe primes such that $p - 1 = 2p'$ and $q - 1 = 2q'$, where p', q' are two primes with length l'-bit. Let $n = pq$ and QR_n be the quadratic residue of Z_n^*. Let h be two generators of QR_n. Also chosen are a group G' of prime order s with length l and two random generators g_1, g_2 of the group G'. We also assume that the discrete logarithm problem is hard in G'.
- Representation of vertex: a vertex $v_i = g_1{}^{x_i} g_2{}^{y_i}$ in an undirect graph G, is an element of group G'.
- Representation of edge: Signature of an edge $\{i, j\}$ is a pair: $\alpha_i = x_i - x_j$ mod s and $\beta_i = y_i - y_j$ mod s in an undirect graph G.
- Certificate of vertex: The certificate of each vertex v_i in authenticated graph is defined by $Cert_i = (e_i, y_i, t_i)$ derived from the signature equation: $y_i{}^{e_i} = Xh^{H(g_1{}^{t_i} g_2{}^{H(v_i)})} \bmod n$.
- A transitive signature: We now can describe our transitive signature scheme: on input 1^k (k stands for the system security parameter), the key generation scheme algorithm creates a pair of signing keys (spk, ssk) for the signature scheme defined above. The signing algorithm $TSign = (VSign, ESign)$ maintains the state of $VSign(i), ESign(i, j)$, where the node $v_i = g_1{}^{x_i} g_2{}^{y_i}$ and a signature of the vertex is defined by $Cert_i = (e_i, y_i, t_i)$ which is derived from the signing equation $y_i{}^{e_i} = Xh^{H(g_1{}^{t_i} g_2{}^{H(v_i)})} \bmod n$. The signature of an edge $\{i, j\}$ is $\delta_{i,j} = (\alpha_{i,j}, \beta_{i,j})$, where $\alpha_{i,j} = x_i - x_j$ mod s and $\beta_{i,j} = y_i - y_j$ mod s.
- The composition algorithm $Comp$: Given nodes v_i, v_j and v_k and the signatures of edge $\{i, j\}$ and edge $\{j, k\}$, it checks the validity of certificate of each node $Cert_i$, $Cert_j$ and $Cert_k$ and it checks the validity of signature of each edge $\delta_{i,j}$ and $\delta_{j,k}$. If all are valid then it outputs $\delta_{i,k} = (\alpha_{i,k}, \beta_{i,k})$.

The undirected transitive signature scheme described above is provably secure under the hardness assumption of strong RSA problem, the hardness assumption of the discrete logarithm problem as well as the H is a collision free hash function in [13].

4.3   Removal of PKI Assumption

We stress that a collection of claimed public keys of a path must be certified. Thus, either a trusted third party (a certificate authority) or a public key infrastructure is required. To remove the concept of certified identity graph G_x from the transitive signatures, we will make use of the following assumption:

each node in an undirected graph G has a unique identity that cannot be forged and it knows the identities of its neighbors in G. We remark that if our keyed-identity graph G_x assumption is not met, an adversary can use different identities to different neighbors. With the help of fixed-identity assumption, an algorithm determining genuine keyed-identity can be proposed. That is, assuming that the underlying graph G is $2k+1$ vertex connected with k adversaries, then between every pair of good nodes, there exists at least $(k+1)$ vertex disjoint paths that traverse only good nodes (the fact that adversaries can at most prove k disjoint paths to a fake node is critical for the solvability of this problem, see [3] and [10] for more details). Based on the above assumptions, we can describe our undirected transitive signature scheme below:

- system parameters: The system chooses a group G' of prime order s with length l and two random generators g_1, g_2 of the group G'. We also assume that the discrete logarithm problem is hard in G'.
- individual system parameters: For each user in the network, it chooses two large safe primes p_i and q_i such that $p_i - 1 = 2p_i'$ and $q_i - 1 = 2q_i'$, where p_i', q_i' are two primes with length l'-bit. Let $n_i = p_i q_i$ and QR_{n_i} be the quadratic residue of $Z_{n_i}^*$. Let X_i and h_i be two random generators of QR_{n_i}.
- representation of vertex: a vertex $v_i = g_1^{x_i} g_2^{y_i}$ in an undirect graph G, is an element of group G'.
- representation of edge: Signature of an edge $\{i, j\}$ is a pair: $\alpha_{i,j} = x_i - x_j$ mod s and $\beta_{i,j} = y_i - y_j$ mod s in an undirect graph G.
- certificate of vertex: The certificate of each vertex v_i in an authenticated graph is defined by $Cert_i = (e_i, z_i, t_i, x_i, y_i)$ which is derived from the equation: $z_i^{e_i} = X_i h_i^{H(g_1^{t_i} g_2^{H(v_i)})} \mod n_i$.

Given a path-finder program, a source node S searches a collection of paths from S to T. Since each node in a graph shares the global system parameters, it follows that each node can be viewed as a self-signed certificate in the transitive graph. Consequently, by applying the technique presented in Section 3, we can calculate the trust value immediately.

5   Conclusion

In this paper, we have introduced and formalized the notion of action in terms of DNF and we have formalized the notion of trust in terms of action, predict function and strategy function. We have already proposed a concise structure for computing of trust value in ad-hoc networks by mapping a path in the underlying network graph G to the corresponding edge of its transitive closure graph G^*. Finally, we have outlined a generic model for computing of trust in ad-hoc networks.

Acknowledgment. The first author is grateful to Professor Herbert Leitold for his invaluable comments on the publication and kind help.

References

1. I.Agudo, J.Lopez, J.A. Montenegro. A Representation Model of Trust Relationships with Delegation Extensions, 3th International Conference on Trust Management (iTRUST'05). LNCS 3477, Springer, 2005.
2. T.Beth, M.Borcherding and B.Klein: Valuation of Trust in Open Networks. ESORICS 1994: 3-18.
3. D.Dolev: The Byzantine Generals Strike Again. J. Algorithms 3(1): 14-30 (1982)
4. J.A. Golbeck. Computing and applying trust in web-based social networks, University of Maryland, College Park, 2005.
5. A.Helmy. Small worlds in wireless networks. IEEE communication letters, Vol.7, No 10, October 2003.
6. A. Jøsang, D. Gollmann, R. Au: A Method for Access Authorisation Through Delegation Networks. Australasian Information Security Workshop 2006.
7. T.Li, H.Zhu, K.Lam: A Novel Two-Level Trust Model for Grid. ICICS 2003: 214-225, Springer Verlag.
8. S.Micali and R.Rivest: Transitive Signature Schemes. CT-RSA 2002: 236-243.
9. M.Reiter and S.Stubblebine. Resilient authentication using path in- dependence. IEEE Transactions on computers, Vol.47, No.12, December 1998.
10. L.Subramanian, R.H.Katz, V.Roth, S.Shenker and I.Stoica: Reliable broadcast in unknown fixed-identity networks. PODC2005: 342- 351.
11. A.C.Yao: Probabilistic Computations: Toward a Unified Measure of Complexity (Extended Abstract) FOCS 1977: 222 -227.
12. R.Yahalom, B.Klein and T.Beth: Trust-Based Navigation in Distribution Systems. Computing Systems 7(1): 45-73, 1994.
13. H.Zhu. New model on undirected transitive signatures. IEE Proceedings of Communication, 2004.
14. H.Zhu, B.Feng and Robert H.Deng. Computing of Trust in Distributed Networks. http:// www.iacr.org, eprint, 2003.
15. H.Zhu, F.Bao and T.Li: Compact Stimulation Mechanism for Routing Discovery Protocols in Civilian Ad-Hoc Networks. Communications and Multimedia Security 2005: 200-209, Springer Verlag.
16. H.Zhu, F.Bao and Robert H.Deng. Computing of Trust in wireless Networks. IEEE Vehicular Technology Conference, 2004.

TAO: Protecting Against Hitlist Worms Using Transparent Address Obfuscation

Spiros Antonatos[1] and Kostas G. Anagnostakis[2]

[1] Distributed Computing Systems Group
Institute of Computer Science
Foundation for Research Technology Hellas, Greece
antonat@ics.forth.gr
[2] Internet Security Lab,
Institute for Infocomm Research
21 Heng Mui Keng Terrace, Singapore
kostas@i2r.a-star.edu.sg

Abstract. Sophisticated worms that use precomputed hitlists of vulnerable targets are especially hard to contain, since they are harder to detect, and spread at rates where even automated defenses may not be able to react in a timely fashion. Recent work has examined a proactive defense mechanism called Network Address Space Randomization (NASR) whose objective is to harden networks specifically against hitlist worms. The idea behind NASR is that hitlist information could be rendered stale if nodes are forced to frequently change their IP addresses. However, the originally proposed DHCP-based implementation may induce passive failures on hosts that change their addresses when connections are still in progress. The risk of such collateral damage also makes it harder to perform address changes at the timescales necessary for containing fast hitlist generators.

In this paper we examine an alternative approach to NASR that allows both more aggressive address changes and also eliminates the problem of connection failures, at the expense of increased implementation and deployment cost. Rather than controlling address changes through a DHCP server, we explore the design and performance of *transparent address obfuscation* (TAO). In TAO, network elements transparently change the *external* address of internal hosts, while ensuring that existing connections on previously used addresses are preserved without any adverse consequences. In this paper we present the TAO approach in more detail and examine its performance.

Keywords: Worms, address space randomization, network security.

1 Introduction

Worms are widely regarded to be a major security threat facing the Internet today. Incidents such as Code Red[1,16] and Slammer[4] have clearly demonstrated that worms can infect tens of thousands of hosts in less than half an

H. Leitold and E. Markatos (Eds.): CMS 2006, LNCS 4237, pp. 12–21, 2006.
© IFIP International Federation for Information Processing 2006

hour, a timescale where human intervention is unlikely to be feasible. More recent research studies have estimated that worms can infect one million hosts in less than two seconds [22,23,24]. Unlike most of the currently known worms that spread by targeting random hosts, these extremely fast worms rely on predetermined lists of vulnerable targets, called *hitlists*, in order to spread efficiently.

The threat of worms and the speed at which they can spread have motivated research in automated worm defense mechanisms. For instance, several recent studies have focused on detecting scanning worms [27,12,26,18,21,25]. These techniques detect scanning activity and either block or throttle further connection attempts. These techniques are unlikely to be effective against hitlist worms, given that hitlist worms do not exhibit the failed-connection feature that scan detection techniques are looking for. To improve the effectiveness of worm detection, several distributed early-warning systems have been proposed [29,17,30,5]. The goal of these systems is to aggregate and analyze information on scanning or other indications of worm activity from different sites. The accuracy of these systems is improved as they have a more "global" picture of suspicious activity. However, these systems are usually slower than local detectors, as they require data collection and correlation among different sites. Thus, both reactive mechanisms and cooperative detection techniques are unlikely to be able to react to an extremely fast hitlist worm in a timely fashion.

Observing this *gap* in the worm defense space, a recent study has considered the question of whether it is possible to develop defenses *specifically* against hitlist worms, and proposed a specific technique called *network address space randomization* (NASR). This technique is primarily inspired by similar efforts for security at the host-level [28,10,9,19,13,8]. It is also similar in principle to the "IP hopping" mechanism in the APOD architecture[7], BBN's DYNAT[14] and Sandia's DYNAT[15] systems, all three designed to confuse targeted attacks by dynamically changing network addresses. In its simplest form, NASR can be implemented by adapting dynamic network address allocation services such as DHCP[11] to *force* more frequent address changes.

The major drawback of the DHCP-based implementation of NASR as presented in [6] is the damage caused in terms of aborted connections. The damage depends on how frequently the address changes occur, whether hosts have active connections that are terminated and whether the applications can recover from the transient connectivity problems caused by an address change. Although the results of [6] suggest that the failure rates are small when measured in comparison to the total number of unaffected connections, the failures may cause significant disruption to specific services that users value a lot more than other connections, such as long-lived remote terminal session (e.g., ssh), etc. Furthermore, the acceptable operating range of DHCP-based NASR does not fully cover the likely spectrum of hitlist generation strategies. In particular, there are likely scenarios that involve very fast, distributed hitlist generation, which cannot be thwarted without extremely aggressive address changes. Aggressive address changes in the DHCP-based NASR implementation have a profound effect on connection failure rates, and the approach hereby becomes less attractive.

As an alternative to the DHCP-based implementation of NASR, in this paper we consider a different approach that allows both more aggressive address changes and also eliminates the problem of connection failures, at the expense of increased implementation and deployment cost. Rather than controlling address changes through a DHCP server, we explore the design and performance of *transparent address obfuscation* (TAO). In TAO, we assume that hosts of a subnet are located behind a network element that transparently changes the *external* addresses of the hosts, while ensuring that existing connections on previously used addresses are preserved without any adverse consequences.

In the rest of this paper, we first discuss in more detail how network address space randomization works generally, and then discuss how transparent address obfuscation can be implemented, and how well it performs.

2   Network Address Space Randomization

The goal of network address space randomization (NASR) as originally proposed in [6] is to force hosts to change their IP addresses frequently enough so that the information gathered in hitlists is rendered stale by the time the worm is unleashed. The authors of [6] have demonstrated that NASR can slow down the worm outbreak, in terms of the time to reach 90% infection, from 5 minutes when no NASR is used to between 24 and 32 minutes when hosts change their addresses very frequently. Their results are based on simulations, varying how fast the hitlist is generated and how fast the host addresses are changed. It appears that the mean time between address changes needs to be 3-5 times less than the time needed to generate the hitlist for the approach to reach around 80% of its maximum effectiveness, while more frequent address changes give diminishing returns. The assumption of global deployment of NASR is unreasonable, thus it is more likely that only a fraction of subnets will employ the mechanism, such as dynamic address pools. NASR continues to be effective in slowing down the worm, even when deployed in 20% or 40% of the network.

The authors of [6] have proposed to implement NASR by configuring the DHCP server to expire DHCP leases at intervals suitable for effective randomization. The DHCP server would normally allow a host to renew the lease if the host issues a request before the lease expires. Thus, forcing addresses changes even when a host requests to renew the lease before it expires requires some minor modifications to the DHCP server. This approach does not require any modifications to the protocol or the client. In their implementation, three timers on the DHCP server for controlling host addresses were used. The *refresh* timer determines the duration of the lease communicated to the client. The client is forced to query the server when the timer expires. The server may or may not decide to renew the lease using the same address. The *soft-change* timer is used internally by the server to specify the interval between address changes, assuming that the flow monitor does not report any activity for the host. A third, *hard-change* timer is used to specify the maximum time that a host is allowed to keep the same address. If this timer expires, the host is forced to change address,

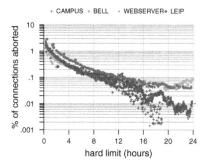
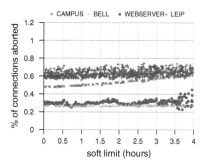

Fig. 1. Percentage of aborted connections as a function of the hard change limit

Fig. 2. Percentage of aborted connections as a function of the soft change limit

as the DHCP server does not renew the lease, despite the damage that may be caused.

The main drawback of this approach is the damage caused in terms of aborted connections. The damage depends on how frequently the address changes occur, whether hosts have active connections that are terminated and whether the applications can recover from the transient connectivity problems caused by an address change. As shown in Figures 1 and 2, the damage varies from 0.01 to 5%. Experiments were done using traces collected at different network environments: a one-week contiguous IP header trace collected at Bell Labs research[2], a 5-day trace from the University of Leipzig[3], a 1-day trace from a local University Campus, and a 20-day trace from a link serving a single Web server at the institute of the authors.

However, as we need to perform randomization in small timescales, where the failure rates wave between 3 and 5%, failure rates may not be acceptable. We can avoid network failures by using *Transparent Address Obfuscation*, an approach which needs more deployment resources than the standard NASR implementation. We describe the *Transparent Address Obfuscation* in the following section.

3   Transparent Address Obfuscation

The damage caused by network address space randomization (NASR) in terms of aborted connections may not be acceptable in some cases. Terminating, for example, a large web transfer or an SSH session would be both irritating and frustrating. Additionally, it would possibly increase network traffic as users or applications may repeat the aborted transfer or try to reconnect. To address these issues, we suggest `Transparent Address Obfuscation`, an external mechanism for deploying NASR avoiding connection failures.

The idea behind the mechanism is the existence of an "address randomization box", called from now on "TAO box", inside the LAN environment. This box

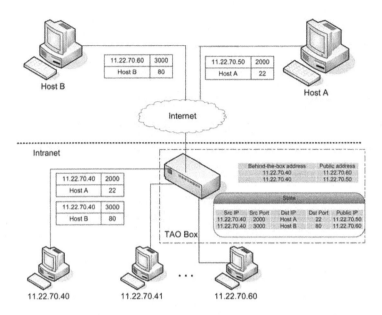

Fig. 3. An advanced example of NASR using the TAO box. Host has two public IP addresses, one (11.22.70.50) devoted for the SSH session to Host A and the other (11.22.70.60) for new connections, such as a HTTP connection to Host B.

performs the randomization on behalf of the end hosts, without the need of any modifications to the DHCP behavior, as suggested in [6]. TAO box controls all traffic passing by the subnet(s) it is responsible for, analogous to the firewall or NAT concept. The address used for communication between the host and the box remains the same. We should note that there is no need for private addresses, unlike the case of NAT, as end hosts can obtain any address from the organization they belong. The public address of the end host – that is the IP that outside world sees – changes periodically according to soft and hard timers, similar to the procedure described in [6]. Old connections continue to operate over the old address, the one that host had before the change, until they are terminated.

The TAO box is responsible for two things. First, to prevent new connections on the old addresses (before randomization) reaching the host. Second, to perform address translation to the packets based on which connection they belong, similar to the NAT case. Until all old connections are terminated, a host would require multiple addresses to be allocated.

An example of how the TAO box works is illustrated in Figure 3. The box is responsible for address randomization on the 11.22.70.0/24 subnet, that is it can pick up addresses only from this subnet. Initially the host has the IP address 11.22.70.40 and TAO box sets the public IP address of this host to 11.22.70.50. The host starts a new SSH connection to Host A and sends packets with its own IP address (11.22.70.40). The box translates the source IP address and replaces it with the public one, setting it to 11.22.70.50. Simultaneously, the box keeps state that the connection from port 2000 to Host A on port 22 belongs to the host with behind-the-box address 11.22.70.40 and public address 11.22.70.50. Thus, on the Host A side we see packets coming from 11.22.70.50. When Host A responds back to 11.22.70.50, box has to perform the reverse translation. Consulting its state, it sees that this connection was initiated by host 11.22.70.40 so it rewrites the destination IP address.

After an interval, the public address of host 11.22.70.40 changes. TAO box now sets its public address to 11.22.70.60. Any connections initiated by external hosts can reach the host through this new public IP address. As it can be seen in Figure 3 the new connection to Host B website has the new public IP as source. Note that in the behind-the-box and public address mapping table host now has two entries, with the top being chosen for new connections. The only connection permitted to communicate with the host at 11.22.70.50 address is the SSH connection from Host A. For each incoming packet, the box checks its state to find an entry. If no entry is found, then packet is not forwarded to the internal hosts, else the "src IP" field of the state is used to forward the packet. As long as the SSH connection lasts, the 11.22.70.50 IP will be bound to the particular host and cannot be assigned to any other internal host. When SSH session finishes, the address will be released. For stateless transport protocols, like UDP or ICMP, only the latest mapping between public and behind-the-box IP address is used.

4 Simulation Study

The drawback of the TAO box is the extra address space required for keeping alive old connections. An excessive requirement of address space would empty the address pool, making the box abort connections. We tried to quantify the amount of extra space needed by simulating the TAO box on top of four traffic traces. The first two traces, CAMPUS and CAMPUS(2), come from a local university campus and include traffic from 760 and 1675 hosts respectively. All hosts of this trace belong to a /16 subnet. The second trace, BELL, is a one-week contiguous IP header trace collected at Bell Labs research with 395 hosts located in a /16 subnet. Finally, the WEBSERVER trace is a 20-day trace from a link serving a single Web server at our institute. In this trace, we have only one host and we assume it is the only host in a /24 subnet. In our simulation, the soft timer had a constant value of 90 seconds, while the hard timer varied from 15 minutes to 24 hours.

The results of the simulation are presented in Figure 4. In almost all cases, we need 1% more address space in order to keep alive the old connections. We measured the number of hosts that are alive in several subnets. We used full TCP scans to identify the number of hosts that were alive in 5 subnets: our local institute, a local University campus and three subnets of a local ISP. Our results, as shown at Figure 5, indicate that 95% of the subnets are less than half-loaded and thus we can safely assume that this 1% of extra space is not an obstacle in the operation of the TAO box. However, the little extra address

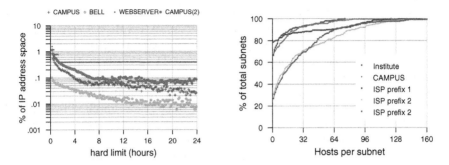

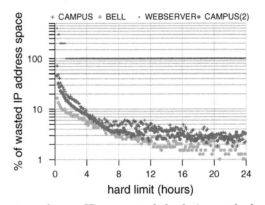

Fig. 4. The percentage of extra IP space needed   **Fig. 5.** Subnet address space utilization

Fig. 6. The percentage of extra IP space needed relative to the load of subnets

space needed derives from the fact that subnets are lightly loaded. For example, the 760 hosts of the CAMPUS trace correspond to the 1.15% of the /16 address space. In Figure 6, the relative results of the previous simulation are shown. On average, 10% more address space for hard timer over one hour is needed, which seems a reasonable overhead. In the case of the WEBSERVER trace the percentage is 100% but this is expected as we have only one host.

5   Related Work

Our work on network address space randomization was inspired by similar techniques for randomization performed at the OS level [28,10,9,19,13,8]. The general principle in randomization schemes is that attacks can be disrupted by reducing the knowledge that the attacker has about the system. For instance, instruction set randomization[13] changes the instruction set opcodes used on each host, so that an attacker cannot inject compiled code using the standard instruction set opcodes. Similarly, address obfuscation[9] changes the locations of functions in a host's address space so that buffer-overflow exploits cannot predict the addresses of the functions they would like to utilize for hijacking control of the system. Our work at the network level is similar, as it reduces the ability of the attacker to build accurate hitlists of vulnerable hosts.

The use of IP address changes as a mechanism to defend against attacks was proposed independently in [7], [14] and [15]. Although these mechanisms are similar to ours, there are several important differences in the threat model as well as the way they are implemented. The main difference is that they focus on targeted attacks, performing address changes to confuse attackers during reconnaissance and planning. Neither project discusses or analyzes the use of such a mechanism for defending against worm attacks.

Reference [20] proposes the use of honeypots with instrumented versions of software services to be protected, coupled with an automated patch-generation facility. This allows for quick (i.e., less than 1 minute) fixing of buffer overflow vulnerabilities, even against zero-day worms. However, that work is limited to worms that use buffer overflows as an infection vector.

While some of these reactive defense proposals may be able to detect the worm, it is unclear whether they can effectively do so in the timescales of hitlist worm propagation.

6   Summary and Concluding Remarks

Fast-spreading malware such as hitlist worms represent a major threat for the Internet, as most reactive defenses currently being investigated are unlikely to be fast enough to respond to such worms in a timely fashion. Recent work on network address space randomization has shown that hitlist worms can be significantly slowed down and exposed to detection if hosts are forced to change their address frequently enough to make the hitlists stale. However, the implications of changing addresses in a DHCP-based implementation, as proposed in [6] hamper the adoption of this defense, as it can cause disruption under normal operation and cannot be performed fast enough to contain advanced hitlist generation strategies.

The approach examined in this paper, Transparent Address Obfuscation (TAO), offers more leeway for administrators to more frequently change addresses, while at the same time eliminating the problem of "collateral damage" in terms of failed connections, when compared to the DHCP-based implementation. The experiments presented in this paper demonstrated that the cost of TAO in terms of additional address space utilization is modest and that the operation of the system is transparent and straightforward.

Acknowledgments

This work was supported in part by the projects CyberScope, EAR and Miltiades, funded by the Greek General Secretariat for Research and Technology under contract numbers PENED 03ED440, USA-022 and 05NON-EU-109 respectively. We also thank Sotiris Ioannidis for his "constructive" comments on an earlier draft of this paper.

References

1. CERT Advisory CA-2001-19: 'Code Red' Worm Exploiting Buffer Overflow in IIS Indexing Service DLL. http://www.cert.org/advisories/CA-2001-19.html, July 2001.
2. NLANR-PMA Traffic Archive: Bell Labs-I trace. http://pma.nlanr.net/Traces/Traces/long/bell/1, 2002.
3. NLANR-PMA Traffic Archive: Leipzig-I trace. http://pma.nlanr.net/Traces/Traces/long/leip/1, 2002.
4. The Spread of the Sapphire/Slammer Worm. http://www.silicondefense.com/research/worms/slammer.php, February 2003.
5. K. G. Anagnostakis, M. B. Greenwald, S. Ioannidis, A. D. Keromytis, and D. Li. A Cooperative Immunization System for an Untrusting Internet. In *Proceedings of the 11th IEEE Internation Conference on Networking (ICON)*, pages 403–408, September/October 2003.
6. Spiros Antonatos, Periklis Akritidis, Evangelos P. Markatos, and Kostas G. Anagnostakis. Defending against Hitlist Worms using Network Address Space Randomization. In *Proceedings of the 3rd ACM Workshop on Rapid Malcode (WORM)*, November 2005.
7. Michael Atighetchi, Partha Pal, Franklin Webber, Rick Schantz, and Chris Jones. Adaptive use of network-centric mechanisms in cyber-defense. In *Proceedings of the 6th IEEE International Symposium on Object-oriented Real-time Distributed Computing*, May 2003.
8. Elena Gabriela Barrantes, David H. Ackley, Trek S. Palmer, Darko Stefanovic, and Dino Dai Zovi. Randomized instruction set emulation to disrupt binary code injection attacks. In *Proceedings of the 10th ACM Conference on Computer and Communications Security*, October 2003.
9. S. Bhatkar, D. DuVarney, and R. Sekar. Address obfuscation: An efficient approach to combat a broad range of memory error exploits. In *In Proceedings of the 12th USENIX Security Symposium*, pages 105–120, August 2003.
10. Jeffrey S. Chase, Henry M. Levy, Michael J. Feeley, and Edward D. Lazowska. Sharing and protection in a single-address-space operating system. *ACM Transactions on Computer Systems*, 12(4):271–307, 1994.
11. R. Droms. Dynamic Host Configuration Protocol. RFC 2131, http://www.rfc-editor.org/, March 1997.
12. J. Jung, V. Paxson, A. W. Berger, and H. Balakrishnan. Fast Portscan Detection Using Sequential Hypothesis Testing. In *Proceedings of the IEEE Symposium on Security and Privacy*, May 2004.
13. Gaurav S. Kc, Angelos D. Keromytis, and Vassilis Prevelakis. Countering Code-Injection Attacks With Instruction-Set Randomization . In *Proceedings of the ACM Computer and Communications Security Conference (CCS)*, pages 272–280, October 2003.
14. Dorene Kewley, John Lowry, Russ Fink, and Mike Dean. Dynamic approaches to thwart adversary intelligence gathering. In *Proceedings of the DARPA Information Survivability Conference and Exposition (DISCEX)*, 2001.

15. John Michalski, Carrie Price, Eric Stanton, Erik Lee Chua, Kuan Seah, Wong Yip Heng, and Tan Chung Pheng. Final Report for the Network Security Mechanisms Utilizing Network Address Translation LDRD Project. Technical Report SAND2002-3613, Sandia National Laboratories, November 2002.
16. D. Moore, C. Shannon, and J. Brown. Code-Red: a case study on the spread and victims of an Internet worm. In *Proceedings of the 2nd Internet Measurement Workshop (IMW)*, pages 273–284, November 2002.
17. D. Nojiri, J. Rowe, and K. Levitt. Cooperative response strategies for large scale attack mitigation. In *Proceedings of the 3rd DARPA Information Survivability Conference and Exposition (DISCEX)*, April 2003.
18. S. E. Schechter, J. Jung, and A. W. Berger. Fast Detection of Scanning Worm Infections. In *Proceedings of the 7ᵗʰ International Symposium on Recent Advances in Intrusion Detection (RAID)*, pages 59–81, October 2004.
19. Hovav Shacham, Matthew Page, Ben Pfaff, Eu-Jin Goh, Nagendra Modadugu, and Dan Boneh. On the effectiveness of address-space randomization. In *CCS '04: Proceedings of the 11th ACM Conference on Computer and Communications Security*, pages 298–307, New York, NY, USA, 2004. ACM Press.
20. S. Sidiroglou and A. D. Keromytis. A Network Worm Vaccine Architecture. In *Proceedings of the IEEE Workshop on Enterprise Technologies: Infrastructure for Collaborative Enterprises (WETICE), Workshop on Enterprise Security*, pages 220–225, June 2003.
21. S. Staniford. Containment of Scanning Worms in Enterprise Networks. *Journal of Computer Security*, 2004.
22. S. Staniford, D. Moore, V. Paxson, and N. Weaver. The top speed of flash worms. In *Proc. ACM CCS WORM*, October 2004.
23. S. Staniford, V. Paxson, and N. Weaver. How to Own the Internet in Your Spare Time. In *Proceedings of the 11th USENIX Security Symposium*, pages 149–167, August 2002.
24. N. Weaver and V. Paxson. A worst-case worm. In *Proc. Third Annual Workshop on Economics and Information Security (WEIS'04)*, May 2004.
25. N. Weaver, S. Staniford, and V. Paxson. Very Fast Containment of Scanning Worms. In *Proceedings of the 13ᵗʰ USENIX Security Symposium*, pages 29–44, August 2004.
26. M. Williamson. Throttling Viruses: Restricting Propagation to Defeat Malicious Mobile Code. Technical Report HPL-2002-172, HP Laboratories Bristol, 2002.
27. Jian Wu, Sarma Vangala, Lixin Gao, and Kevin Kwiat. An Effective Architecture and Algorithm for Detecting Worms with Various Scan Techniques. In *Proceedings of the Network and Distributed System Security Symposium (NDSS)*, pages 143–156, February 2004.
28. J. Xu, Z. Kalbarczyk, and R. Iyer. Transparent runtime randomization for security. In *A. Fantechi, editor, Proc. 22nd Symp. on Reliable Distributed Systems –SRDS 2003*, pages 260–269, October 2003.
29. Vinod Yegneswaran, Paul Barford, and Somesh Jha. Global Intrusion Detection in the DOMINO Overlay System. In *Proceedings of the Network and Distributed System Security Symposium (NDSS)*, February 2004.
30. C. C. Zou, L. Gao, W. Gong, and D. Towsley. Monitoring and Early Warning for Internet Worms. In *Proceedings of the 10ᵗʰ ACM International Conference on Computer and Communications Security (CCS)*, pages 190–199, October 2003.

On the Privacy Risks of Publishing Anonymized IP Network Traces

D. Koukis[1], S. Antonatos[1], and K.G. Anagnostakis[2]

[1] Distributed Computing Systems Group, FORTH-ICS, Greece
{koukis, antonat}@ics.forth.gr
[2] Infocomm Security Department, Institute for Infocomm Research, Singapore
kostas@i2r.a-star.edu.sg

Abstract. Networking researchers and engineers rely on network packet traces for understanding network behavior, developing models, and evaluating network performance. Although the bulk of published packet traces implement a form of address anonymization to hide sensitive information, it has been unclear if such anonymization techniques are sufficient to address the privacy concerns of users and organizations.

In this paper we attempt to quantify the risks of publishing anonymized packet traces. In particular, we examine whether statistical identification techniques can be used to uncover the identities of users and their surfing activities from anonymized packet traces. Our results show that such techniques can be used by any Web server that is itself present in the packet trace and has sufficient resources to map out and keep track of the content of popular Web sites to obtain information on the network-wide browsing behavior of its clients. Furthermore, we discuss how scan sequences identified in the trace can easily reveal the mapping from anonymized to real IP addresses.

1 Introduction

Packet-level traces of Internet traffic are widely used in experimental networking research, and have been proved valuable towards understanding network performance and improving network protocols (c.f. [18,13,11,12]). Since raw packet traces contain sensitive information such as emails, chat conversations, and Web browsing habits, organizations publishing packet traces employ techniques that remove sensitive information before making the traces available.

The most common process of "anonymizing" a packet trace involves removing packet payloads and replacing source and destination IP addresses with anonymous identifiers [14]. Several repositories of trace datasets have been made available that employ this technique and they have served the research community extremely well for many years[2,3]. Without payloads and host IP addresses, it is widely assumed that it is hard, if not impossible, to elicit any sensitive information from a packet trace. However, several researchers have recently expressed concerns, albeit without further investigation, that there *may* be ways of indirectly *inferring* sensitive information from sanitized traces[20,16]. This kind of

H. Leitold and E. Markatos (Eds.): CMS 2006, LNCS 4237, pp. 22–32, 2006.

information on individual users and organizations could be used in many ways that may be illegal or simply in a bad way (e.g., for marketing, surveillance, censorship, industrial espionage, etc.).

In this paper, we attempt to experimentally assess the risk of publishing "anonymized" packet traces. In particular, we examine whether it is possible to break the address anonymization scheme and identify specific host addresses in the network traffic packet trace. We focus on two potential statistical identification attacks: one using known web site footprints, and one using known patterns of port-scanning activity. The first attack has the potential of discovering the anonymized identifiers of known Web servers, which can be linked to (still anonymous) client IPs. The second attack is not restricted to a specific type of host or protocol, and is therefore "ideal" for recovering client IPs which cannot be recovered using the first technique alone.

2   Related Work

Although our work is not the first to raise questions about the risks of publishing anonymized packet traces, to the best of our knowledge it is the first report that tries to provide some answers. The issue was first raised by Ylonen[20], who discussed several ways of breaching the privacy of traces that have been anonymized using *tcpdpriv*[14]. This report also mentions matching known website footprints to the packet trace. The same attack is also imagined by Pang and Paxson in [16]. In both cases, the authors did not further examine the technical details of the attack and did not experimentally quantify its potential effectiveness.

Launching a "known-plaintext" attack on web-site footprints found in packet traces is similar in many ways to launching such an attack on footprints found in encrypted Web traffic, as seen by a Web proxy. For the case of the proxy server, Sun et al.[19] developed a technique that monitors a link between proxy and clients, gathers HTTP requests and responses, and compares them against a database of signatures. The authors show that most of the Web pages in their database can be identified and that false positives are rare. Similar results are presented by Hintz in [10]. Moreover, Pang et al. in [17] mention the problem of sequential scans found in traces that could lead to a potential disclosure of the anonymized IP addresses.

Although the basic idea of launching a known-plaintext attack on encrypted Web proxy traffic is similar to attacking a packet trace, there are two major differences that justify further investigation. First, attacking packet traces instead of proxy traffic seems a lot easier and thus more threatening, as anonymized packet traces are widely available through organizations such as NLANR[4] In contrast, web proxy logs are usually not shared outside an organization. Thus, the technique examined in this paper is an existent threat to the privacy of users.

Second, it seems difficult to directly apply the proxy technique to packet traces. We have identified several issues that need to be addressed in attacking a packet trace that were not explored in the proxy attack. For instance, caching

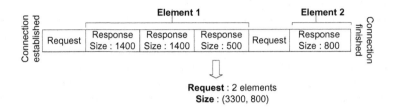

Request : 2 elements
Size : (3300, 800)

Fig. 1. Extracting elements from a packet trace. The responses between two successive requests belong to a single element (Element 1) and the element size is the sum of response packets. The connection termination also indicates the end of an element (Element 2).

schemes, the use of cookies, the type of browser used to make the request and various HTTP options and protocol details alter the underlying input and may influence the effectiveness of the method. Another problem that we faced in our analysis is the reconstruction of the HTTP data from the TCP/IP traces. This was not the case in Web proxy traffic where there is no need for HTTP-level reconstruction as the complete request sequence for a page is available. This issue affects the matching process and should be taken into account.

3   Identification of Anonymized Web Server Addresses

3.1   Extracting Web Signatures

Before describing the process for extracting HTTP requests from packet traces, it is important to understand exactly how the protocol under attack is implemented. For HTTP protocol version 1.0[7], a new connection should be created with the web server in order to retrieve a new element. The term element defines every item that contitutes a web page, which may include html pages, images, css and script files and in general everything that can be referenced in a web page. To reduce the delay perceived by the client during web browsing and additionally the traffic produced, HTTP version 1.1[9] introduced *persistent connections*. A persistent connection does not need to be terminated after retrieval of an element but can be reused to process further requests to the same server. Current browsers use HTTP/1.1 by default, although the user can specify to use HTTP 1.0. Typically, when a browser requests an HTML page, the retrieved page is parsed in order to find its embedded objects. Afterwards, an HTTP request is issued for each one of these objects. Responses sent by the server consist of the HTTP headers along the actual data that were requested.

The first step for signature extraction is to identify web traffic from traces. This is is relatively easy by taking packets to and from port 80. Packets with zero payload size are considered as acknowledgments. The second step is to analyze HTTP responses. In case of HTTP/1.0, a request is made, the response follows and then connection terminates. Thus, the size of the element is the sum of payload sizes of packets with source port 80 (for this flow). In case of HTTP/1.1 (default for most browsers), we may have multiple requests on the same connection. All the packets from server to client that are found between two requests belong to the same response and this response belongs to the first request. The process is illustrated in Figure 1. In this study we do not consider "HTTP pipelining" – this would somewhat complicate our analysis, but is outside the scope of this work, since it is still an experimental feature, disabled by default in all browsers, and rarely used.

Having collected a set of elements, we need to assign them to the web page they belong. Elements are grouped on a per-client basis. It is necessary to distinguish successive Web page requests and assign the identified elements to the right page. A web page may contain elements from multiple web servers. Often, HTML files reside on the main web server, while images or embedded objects (like ads or banners) are loaded from another server. This behavior does not allow us to rely on IP address for assigning elements. Another problem arises when there are two requests for the same web server, e.g., when a user clicks a hyperlink for a page on the same domain as the original web page. Given that we have no HTTP-level information, we have to rely on a heuristically determined timeout value. After a user downloads a page, we assume that there is an "idle" period until the user clicks a link to another page. Figure 2 shows the process of assignment.

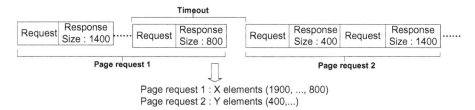

Fig. 2. Assigning elements into page requests. The timeout value is used to separate requests for two different pages.

After the assignment of elements to Web pages, a signature for that page can be created. As we have no payload from traces, the element size is the most promising piece of information that we can extract. Note that the size of each element is computed including the HTTP response headers, which causes variation of element size even for subsequent requests (due to variable length fields like cookies and date). Consequently, the signature for a web page is the set of element sizes that is consisted of. This definition for web page signature creation has several advantages but also limitations. As the number of elements forming a page increases, signatures become more accurate. However, it is possible for a web page to generate more than one signature for three main reasons. First, the web page may be dynamic or changing periodically. Dynamic pages may alter the actual content of the page (HTML file) and the embedded objects, for example, to present different ads each time. Second, due to HTTP headers length variation, even for static pages we may get different signatures. Finally, caching may affect the creation of the signature and provide different results for the same page.

3.2 Web Server Fingerprinting Methodology

A fingerprint attack can be formed as follows. The first step is to obtain a signature for each target page that is to be identified. As mentioned before, a signature for a page may not be constant and thus, using information just from one signature may reduce the probability of a successful match. It is possible to obtain many signatures for the same page and extract all the elements into a unified set. The members of this set are most likely to be found in the trace as a subset of a complete request. Additionally, in order to compensate for minor HTTP header variations that may occur, a minor padding (ranging from 8 to 32 bytes) could be applied to the size of the elements. We should note here that signatures should have been created at approximately the same time as the trace packets because even static web pages may change over time. Moreover, if a target site is dynamic, multiple signatures should be created in order to match all possible appearances of the page.

After the signature database has been created, HTTP elements should be extracted from the packet trace and grouped into page requests. Since the adversary has no way to find out whether HTTP reconstruction was successful and up to which degree, this step can be done multiple times using different timeout values.

The next step is the matching process. Information extracted from the trace should be compared against the signatures contained in the database. The elements extracted from the trace are compared against the elements of each signature and a similarity score is computed. The similarity score is the percentage of common elements between the request on the trace and the signature. If the score is above a threshold then we have a potential match. As each web site has multiple signatures, if we have more than one matche for a site, our guess is more confident. If the request on the trace matches multiple web sites, then the site that gives the highest scores is selected as a possible match. The complete fingerprinting methodology is shown in Figure 3.

3.3 Experimental Results

A key parameter of the fingerprinting process is how signatures change over time. Initially, ten thousand different pages were collected from Google's web directory [1], which was choosed due to the large number of indexed pages. Pages were chosen randomly from various categories and only a few of them belong to popular web servers. For each page, its signature was generated multiple times in different time periods.

In our first experiment, we examined whether collected signatures remain constant over time. For each target site, we created its signature every half an hour. We target to calculate the percentage of static pages in the web, as an indication of whether the fingerprinting method can be applied succesfully.

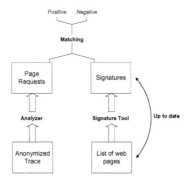

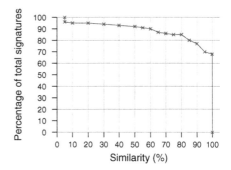

Fig. 3. Fingerprinting attack methodology

Fig. 4. Similarity among signatures

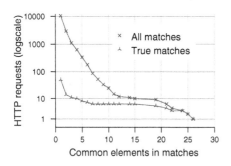

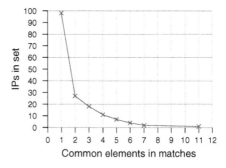

Fig. 5. Number of HTTP requests that share common objects with those found in the signature of target page

Fig. 6. Number of distinct IP addresses that exist in matches and share common elements with the target signature

Previous works [8] show that about 56% of the total pages appear to be static. In Figure 4, the similarity percentage for different signatures of the same page is shown. It can be observed that 67% of page signatures remain constant, while 90% of them have more than 60% similarity. It is clear that there still exist a large portion of static pages that give potential for our method to work.

In our second experiment, requests found in a trace collected at a local subnet of our institute were compared against the signature of the main page for a popular web server. The signature contained 27 elements and we had in our disposal both the anonymized and non-anonymized form of the trace to verify our results and identify true matches. In Figure 5, the number of requests that match the signature as a function of the number of common elements is presented. As we increase the number of common elements, the number of requests that match the signature decreases. Furthermore, near the maximum value of common elements (27) false positive ratio, e.g. potential matches against true matches, drops to zero.

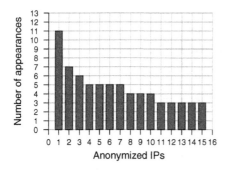
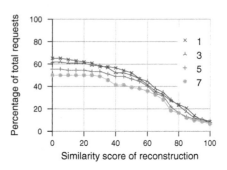

Fig. 7. Histogram of anonymized IP addresses found in requests with common elements to the target signature

Fig. 8. Efficiency of the reconstruction process for assigning elements into sets

It is possible that the algorithm does not always succeed to come up with one result for a signature. Instead, two or more matches may contain the same number of common elements with multiple signatures. In this case the true match can be decided based on the number of times that each match has been found. This way, we try to find many matches to the target signature that each one has a few common elements. The first step of this process is to group requests into sets based on the number of common elements with the target signature and count distinct IP addresses found in these requests. Figure 6 shows the number of IP addresses in each of these sets. When the number of common elements reaches its maximum value the number of IP addresses remains almost constant so there is a precise guess.

In Figure 7, the histogram of the fifteen most popular IP addresses in the matches is shown. The IP address IP1 is by far the most frequently appearing address (considering that the number of web requests in the trace is limited to few). After evaluating this experiment with the non-anonymized trace, we have found that this IP was the right mapping for the target site.

In order to improve confidence of the guess, we could correlate the results of the previous methods. If both methods end up to the same result, that means identifying the IP address which has more common elements than all other requests and also is the most popular, our guess can be more confident.

As the techniques described previously depend on the information extracted from traces, we evaluated the HTTP reconstruction process. More precisely, we measured the correctness of assigning elements into Web pages. We analyzed the non-anonymized trace using HTTP level information and extracted the Web page along with their elements. Then the algorithm of assignment was applied to the anonymized trace for multiple timeout values. In Figure 8, we present the percentage of requests found in trace that have been reconstructed correctly, for timeout values 1, 3, 5 and 7 seconds. The similarity score of the reconstruction is computed as the percentage of elements of the request that have been identified using our technique compared to the number of real elements. As we can see

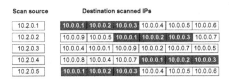

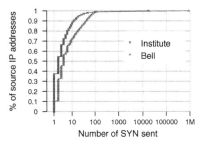

Fig. 9. Passive scan-based identification tries to identify the MCS in the SYN packets (marked as red)

Fig. 10. CDF of SYN packets sent by hosts in BELL trace and in a trace captured at our institute

about 8% of the requests can be correctly reconstructed and this is common for all timeout values. Although this is a rather low percentage, the existence of these requests is an evidence that matching can be partially successful. Also, requests that remain constant regardless the timeout value are more probable to have been correctly reconstructed and matching should be based on these.

4 Passive Scan-Based Identification

It is trivial for an attacker to discover the real IP addresses from an anonymized trace if he can inject a scan to a specific IP address range. Given that the attacker knows the range that is scanned, and assuming he is able to identify his own packets in the trace, obtaining the mapping of any anonymized IP address to the real address is straightforward. However, such attempts may also be easy to detect, hereby exposing the attacker. In some cases, the attacker may not be willing to take such a risk. In this Section we present a more stealthy attack that does not require the attacker to inject any traffic in the trace.

Instead of relying on injected scans, our algorithm relies on passive identification and analysis of existing scans performed by others, using tools such as nmap[5]. Scans found in packet traces have different forms: some are linear scans targeting specific subnets, others are random scans within the same subnet, others are completely randomly generated IP addresses throughout the Internet. The number of elements in a scanning sequence may also vary, although it is not uncommon for such tools to map whole subnets.

To illustrate the basic operation of our algorithm we first describe a simple scenario. Assume a target /24 subnet a.b.c.X/24 for which the anonymized IPs need to be mapped to real IPs, and a trace that contains only full scans across the subnet. The scans are either linear or random. If there are at least two scan sequences that are identical, we know that with very high probability these scans are sequential. The sequential scans then directly provide us with the mapping of the /24 subnet, as the sequence identified in the anonymized trace can be mapped to addresses a.b.c.1 through a.b.c.254.

This observation can be generalized to any set of scan sequences. We assume that some of the sequences are in random order and some of the sequences are linear. Our goal is to identify maximum common subsequences between different scans. The longer the common subsequence, the more likely it is that it represents a linear scan. Since short sequences are more likely to be coincidental, we cannot simply consider every pairwise match as legitimate. Instead, we iteratively construct the complete map of a subnet from pairwise matches in order of confidence, and look for the maximum consistent set of pairwise matches that cover the whole subnet.

To evaluate our approach, we used a network trace collected from two /24 subnets, locally at our institution. The duration of the trace was 4 days and it contained header-only information. We had at our disposal both the anonymized and non-anonymized version. The first step is to recognize the source IP addresses that perform scanning. A simple heuristic is to select only these hosts that send a large number of SYN packets. We measured the cumulative distribution function of SYN packets sent by hosts in both our trace and a network trace from Bell Labs which had the same duration with our trace. The results are summarized in Figure 10. It can be observed that only 1% of the hosts generate more than 80 SYN packets in the whole duration of the trace, thus leaving us only a small percentage of hosts to be investigated. For our next experiment, we used 80 as the threshold for selecting hosts that are considered as scanners.

After having selected the set of source IP addresses to check, we try to find the longest subsequence of destination hosts that they sent SYN packets to. We found on the trace two IP addresses that shared a subsequence of 512 destination hosts. After looking at the non-anonymized traces we verified that these two IP addresses were scanning linearly a local subnet, sending two SYN packets per host, apparently using the nmap tool. Although this is a specific example with a large common subsequence, it demonstrates the effectiveness of our technique as it is indicative of how our approach would work on real traffic. In lack of a reasonably large trace with both anonymized and real addresses, we were unable to evaluate in more detal the identification of linear scans through smaller common subsequences where the probability of false matches is higher. However, we believe that long linear scans are very likely to occur even within shorter timescales, thus enabling the straightforward application of our algorithm with high confidence.

5   Concluding Remarks

In this paper we have examined whether an adversary can break the privacy of anonymized network packet traces. Such a threat is significant, as there are organizations that publish packet traces from their networks for research purposes.

We have examined two attacks: one that identifies web servers in the trace through known web site fingerprints, and one that attempts to recover the original IP addresses in the trace from well-structured port-scanning activity. Associating the inferred information with other sources, such as web server's log files, could lead to significant privacy problems. Our results show that these attacks are reasonably effective, despite being inexact and error-prone.

Although our results are not particularly surprising, they provide solid experimental evidence confirming the concerns previously expressed by other researchers. One interesting observation is that the attack may be more complex and error-prone than previously thought. However, it is not unlikely that careful engineering can lead to higher identification accuracy compared to our results.

Since real-world datasets are essential for research, our results also reinforce the need for alternatives to publishing sanitized packet traces. One possible direction is the use of systems that have access to the raw traces but only allow access to them through a query interface[6,15]. Since the system controls the queries, it may be possible to control their privacy implications, or at least retain an audit trail in case a privacy violation is discovered at a later point.

Acknowledgments

This work was supported in part by the IST project LOBSTER funded by the Europen Union under Contract No. 004336 and the GSRT project Cyberscope funded by the Greek Secretariat for Research and Technology under the Contract No. PENED 03ED440. We would also like to thank Kostas Anagnostakis for his insightful comments. Spiros Antonatos is also with the University of Crete.

References

1. Google's directory. http://directory.google.com.
2. The internet traffic archive. http://www.acm.org/sigs/sigcomm/ITA.
3. NLANR network traffic packet header traces. http://pma.nlanr.net/Traces/.
4. Nlanr passive measurement and analysis. http://pma.nlanr.net/PMA/.
5. Remote OS detection via TCP/IP Stack FingerPrinting. http://www.insecure.org/nmap/nmap-fingerprinting-article.html, June 2002.
6. K. G. Anagnostakis, S. Ioannidis, S. Miltchev, J. Ioannidis, Michael B. Greenwald, and J. M. Smith. Efficient packet monitoring for network management. In *Proceedings of the 8th IEEE/IFIP Network Operations and Management Symposium (NOMS)*, pages 423–436, April 2002.
7. T. Berners-Lee, R. Fielding, and H. Frystyk. RFC 1945: Hypertext Transfer Protocol — HTTP/1.0, May 1996.
8. B. E. Brewington and G. Cybenko. How dynamic is the Web? *Computer Networks (Amsterdam, Netherlands: 1999)*, 33(1–6):257–276, 2000.
9. R. Fielding, J. Gettys, J. Mogul, H. Nielsen, and T. Berners-Lee. Hypertext transfer protocol - HTTP/1.1. *RFC 2616*, June 1999.
10. A. Hintz. Fingerprinting websites using traffic analysis. In *Privacy Enhancing Technologies (PET 2002)*, pages 171–178. Springer-Verlag, LNCS 2482, 2002.
11. H. Jiang and C. Dovrolis. Passive estimation of tcp round-trip times. *Computer Communications Review*, July 2002.
12. S. Jin and A. Bestavros. Sources and characteristics of web temporal locality. In *MASCOTS*, pages 28–35, 2000.

13. M. Mathis, J. Semke, J. Mahdavi, and T. Ott. The macroscopic behavior of the TCP congestion avoidance algorithm. *ACM Computer Communication Review*, 27(3), July 1997.
14. G. Minshall. Tcpdpriv: Program for eliminating confidential information from traces, 2005. http://ita.ee.lbl.gov/html/contrib/tcpdpriv.html.
15. J. Mogul. Trace anonymization misses the point. Presentation on WWW 2002 Panel on Web Measurements.
16. R. Pang and V. Paxson. A High-Level Programming Environment for Packet Trace Anonymization and Transformation. In *Proceedings of the ACM SIGCOMM Conference*, August 2003.
17. Ruoming Pang, Mark Allman, Vern Paxson, and Jason Lee. The devil and packet trace anonymization, January 2006.
18. V. Paxson and S. Floyd. Wide-area traffic: the failure of Poisson modeling. In *Proceedings of ACM SIGCOMM*, pages 257–268. August 1994.
19. Q. Sun, D. R. Simon, Y. Wang, W. Russell, V. N. Padmanabhan, and L. Qiu. Statistical identification of encrypted web browsing traffic. In *Proceedings of IEEE Symposium on Security and Privacy*, Oakland,CA, May 2002.
20. T. Ylonen. Thoughts on how to mount an attack on tcpdprivs "-a50" option. http://ita.ee.lbl.gov/html/contrib/attack50/attack50.html.

Secure Mobile Notifications of
Civilians in Case of a Disaster

Heiko Rossnagel and Tobias Scherner

Chair of Mobile Commerce and Multilateral Security
Johann Wolfgang Goethe - University Frankfurt,
Gräfstr. 78, 60054 Frankfurt, Germany
heiko.rossnagel@m-lehrstuhl.de,
tobias.scherner@m-lehrstuhl.de
http://www.m-lehrstuhl.de

Abstract. Disaster management using mobile telecommunication networks provides a new and attractive possibility to save human lives in emergencies. With this contribution, we present a possible disaster management system based on mobile telecommunication. In order to use such a system in the real world, security requirements such as availability, accountability, integrity and confidentiality have to be ensured by the disaster management system (DMS). We summarize these requirements and propose ways of addressing them with a multilateral secure approach. Using electronic signatures based on SIM-cards, we assure integrity, accountability and confidentiality of the notification messages. We also discuss how availability could be increased.

1 Introduction

Historic examples demonstrate that disasters had a strong influence on the development of nations [3]. Sometimes, the consequences were so enormous that the effected cultures vanished from the world's stage [26].

People usually have problems to recognize leading signs of natural disasters and the possible magnitude of damages. One example is the eruption of Mount Thera and the disappearance of the Minoan culture [17] [7]. Another tragic example is the 2004 Tsunami [34].

Disasters can be caused by natural reasons or can be driven by humans. In the majority of cases, the latter ones do not have any leading signs, like for example the Chernobyl explosion [35] or the September 11th attacks [33]. Therefore, promptly notification and evacuation of the people who are endangered by the disaster is especially desirable in order to save as much lives as possible.

Mobile communication infrastructures offer standardized wireless communication services in almost all countries [19] and allow a fast diffusion of information. This existing and deployed infrastructure could be used for emergency service applications using location-based services (LBS). Currently, these emergency services are discussed and standardized by organizations and bodies like the European Telecommunications Standards Institute (ETSI) with the aim of preparing a framework for worldwide interoperable emergency services [15] [13] [12] [14]. In

H. Leitold and E. Markatos (Eds.): CMS 2006, LNCS 4237, pp. 33–42, 2006.
© IFIP International Federation for Information Processing 2006

addition, the European Commission is strongly interested in this topic and encourages research and standardization of electronic communication networks [9] [25]. Thereby, the European approach focuses on what can be delivered instead of defining services and service levels without having the available technology like the E911 project [5]. Along with these new opportunities of fine-grained disaster management, new possibilities of abuse also do emerge. For example, it is possible to send fake disaster warnings via the Short Message Service (SMS) [27]. Therefore, it is necessary to analyze the security requirements of such DMSs and to meet these requirements when designing a future system. Naturally, privacy concerns have to be discussed in the setting of mobile network based DMS. However, this is out of scope of this paper as privacy issues have already been discussed in [18].

In section 2 of this paper, we present a DMS based on mobile communications infrastructure similar to [30] [18]. We then analyze the security requirements of such a system in section 3, propose some refinements in order to address the requirements in section 4 and then conclude our findings.

2   Disaster Management System

DMS are complex systems and should be designed in an integrated approach from detecting events up to eliminating possible threats to people and infrastructures [16].

In particular, DMSs should enable disaster forces to manage disaster events, including detection and analysis of incidents. Persons in charge should be supported to prepare evacuations, control and support disaster forces and to locate victims. An example of requirements with local characteristics of a DMS in Indonesia can be found in [2].

2.1   General Requirements

Yuan and Detlor [36] have undertaken a possible categorization of requirements. Based on this study, Scherner and Fritsch [30] augmented this categorization by extending it to popular and promising technologies that are currently in use or being discussed, and analyzing their strengths and weaknesses. Their analysis shows that mobile communication infrastructures are superior to other technologies. Some of these advantages are:

- Identification and locating experts
- Custom tailored messages to different parties and locations
- Dynamical notification updates while individuals are passing over to another danger zone
- Measuring of movements of the holders of mobile phones
- Providing back channels to victims[30]

However, to use this technology, the market penetration of mobile devices and the network coverage have to be sufficient. Both factors are crucial for success of a mobile network-based DMS. Currently, worldwide over 1.5 Billion GSM-subscribers are registered [1]. The market penetration of mobile devices differs in Western Europe between 97,1% in Sweden and 68,8% in France (population / mobile subscribers[1]) [6]. Naturally, it is impossible to make general statements about the network coverage in Europe. Multiple factors have influenced the development of mobile networks in different countries. Examples are the amount of fixed lines before the emergence of mobile networks, the population density, and economic drivers that speed up different communication technologies.

Nevertheless, even sparsely populated countries have invested in mobile networks, instead of providing fixed-lines in remote areas. One example is the GSM-network coverage of Sweden, which is shown in Fig. 1.

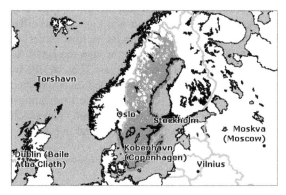

Fig. 1. GSM-network coverage of Sweden [18]

2.2 A DMS Based on GSM Networks

The participating parties in our scenario are mobile subscribers, disaster managers, mobile operators, and a DMS. Mobile subscribers are able to register themselves during the preparatory phase (before the occurrence of disasters) and can define and approve observation rules. The observation rules are stored in a separated part of the infrastructure and are executed if one of the parties is located within a defined disaster area. In Addition, geographical areas can also be observed. An exemplary use case is the observation of chemical warehouses by safety inspectors.

Furthermore, users are able to register themselves as specialists like medics, fire fighters or other disaster forces. This self-declaration as a specialist has to be confirmed by the employer or aid organization. Because instructions for specialist are tailored for the individual recipient, these messages and the replies to disaster managers have to be sent by point-to-point technologies like SMS or Multimedia Message Service (MMS). On possible restrictions on applying point-to-point technologies in emergency cases, see [22] and [8]. In contrast, warnings to civilians will be send via Cell Broadcast Service (CBS). CBS belongs to the point-to-multipoint technologies and offers the following useful characteristics [23]:

- CBS has very low setup costs for operators, users and disaster managers.
- Activation of CBS can be provided by the operator via SIM Application Toolkit (SIM AT).

---

[1] A certain subscriber is counted multiple times if he has several mobile communication accounts.

- CBS reduces the traffic as recipients in the disaster area receive the notification just in time.
- Privacy concerns about CBS do not exist.
- Mobile networks can be secured against power outages.
- Mobile phones offer the possibility of direct communication between rescue forces and victims.

Disaster managers use a geographic information system (GIS)-supported platform to manage disaster activities, like warnings, locating and routing of victims, and controlling the disaster forces as described in [18] and similar in [37]. If a disaster event occurs, the disaster manager sends out warnings to the effected areas by cell broadcast to ensure in-time warnings of potential victims. Afterwards, he is able to locate the victims and pre-registered specialists through the DMS. This information is required for controlled evacuations of disaster areas. Thereby, the timing of warnings in different areas can be used to prevent overcrowded escape routes. The accuracy of the detected positions may differ from cell to cell due to locating methods and cell dimensions [37]. This has to be considered while planning and executing evacuations. An overview of the proposed infrastructure can be found in Fig. 2.

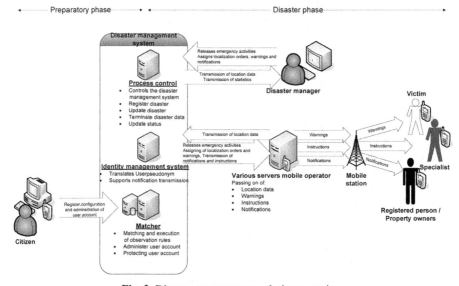

Fig. 2. Disaster management solution overview

Mobile operators provide the communication infrastructure, send out warnings and deliver location information based on cell IDs.

The DMS is the core component of our proposed infrastructure and consists of a middleware solution between mobile operator and disaster manager. It is separated into three independent architectural elements, called Matcher, Identity Management Control (IDM), and Process Control. The Matcher locates civilians within the disaster area. It matches the disaster area with observation rules of the users and protects persistent store of individual observation rules. It matches profiles of threatened person with their registered contact person. The Identity Management System controls the information exchange between the disaster management, the mobile operator and the civilians. Information exchange between disaster manager and mobile operator is done by using different user pseudonyms to avoid linkability for unauthorized observations. Borking [4] described this kind of identity protector first. The Process Control administers the DMS, represents an interface to the disaster manager and is responsible for temporary storage of disaster data (observation rules and localization information).

Advantages of this system are the ability to monitoring victims, pseudonymous identification of certain mobile subscribers, and operational control of individuals.

3 Security Requirements of the Notification Infrastructure

The proposed infrastructure also has to be protected in regard to the four traditional security targets availability, integrity, accountability and confidentiality [28]. Therefore, we develop a set of application and security requirements in this section that has to be fulfilled by the DMS.

3.1 Notification of Mobile Subscribers

In order to leave as much time as possible for evacuation, the notification of civilians has to be as little time consuming as possible. Therefore, we formulate an application requirement.

Requirement I:	*The notification process has to be as little time consuming as possible.*

Since mobile subscribers use a great variety of different mobile devices, the service should be as compatible as possible to most devices. Therefore, we formulate another requirement.

Requirement II:	*The notification service should be useable with (almost) any mobile phone on the market.*

Because of the short time span in which the mobile subscriber has to react to the notification it is not possible to crosscheck the received information. Therefore, the notification service must provide a way to ensure the integrity and authenticity of the notifications. Otherwise, a potential attacker could alter notification messages or create false notification messages that could lead to a disaster by itself. Terrorists for example could use the DMS to create a mass panic.

Requirement III:	*The notification service has to ensure that the user can determine that the notification is from an authorized disaster management authority and that the integrity of the notification message has not been violated.*

It is also important that the notification does not get lost or delayed (availability is the corresponding property). Consequently, we formulate a third requirement, which should be fulfilled by the notification service. Obviously, the notification service on its own cannot guarantee fulfillment of any of these requirements (e.g. when communications are interrupted or tampered within parts of the network outside of its control), but it is important that the user knows about the state of the message he receives.

Requirement IV::	*The notification system has to ensure that messages reach subscribers in time.*

3.2 Notification of Specialists

Since specialists are a special form of mobile subscribers all requirements stated above also apply for them. Furthermore, some additional requirements have to be determined for specialists. If the DMS is notifying specialists, there might be reasons that some of this information should not be publicly available. Therefore, the notification service should, in addition to the requirements stated above, provide means to ensure that confidentiality of the notification is preserved. Therefore, we formulate another requirement.

Requirement V:	*When notifying specialists, the notification messages should be confidential.*

Furthermore, the specialist should be capable to interact with the disaster manager, in order to provide updates of the local situation and for instance, his availability to ease up resource scheduling. Therefore, the specialist will send messages back to the disaster manager. However, the disaster manager has to be able to verify the authenticity and integrity of these messages. Otherwise, a potential attacker could change messages or create false ones in order to hamper rescue efforts. In addition, these messages should also be confidential. Therefore, we present two more requirements.

Requirement VI:	*Specialists have to be able to send confidential messages to the disaster manager.*

Requirement VII:	*Disaster managers have to be able to check the authenticity and integrity of incoming messages.*

4 Further Refinement of Proposed Infrastructure

The changes to the infrastructure we are going to propose are based on the assumption that the mobile subscribers are using SIM cards that are capable of creating and verifying electronic signatures. The technology for such SIM cards exists but has not gained much market penetration so far. The WiTness project [10] sponsored by the

European Union has developed such a SIM card that is capable of creating RSA signatures [29] and also provides 3DES encryption. Using such a SIM card, the mobile subscriber can obtain a copy of the public key of the notification service provider. Furthermore, specialist can register their public keys in the DMS. Having defined the necessary premises, we can now propose the following infrastructure that is illustrated in Fig. 3.

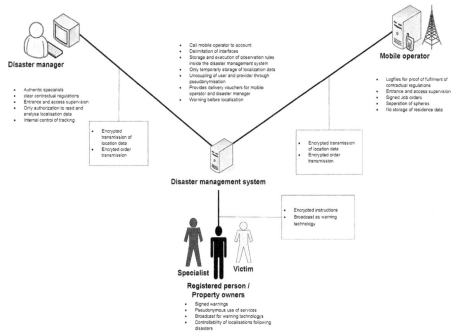

Fig. 3. Notification system infrastructure (proposed)

Our goal is to achieve as many of the application and security requirements defined in section 3 as possible. Therefore, we propose an implementation using the SIM AT, which also ensures compatibility to almost all mobile phones.

When a disaster occurs, the disaster manager initiates a notification and sends it to the SIM AT application running on the mobile device of the addressee [11] by SMS in case of specialists or CBS for normal subscribers. The notification is electronically signed with the private key of the notification service provider. After receiving the signed push notification, the application can check the integrity and authenticity of the notification by verifying the signature. If the signature is valid, the mobile subscriber can now follow the given instructions. Since they have registered their public keys, specialists are also able to send electronically signed messages back to the disaster manager, who can check the authenticity and integrity of these messages. Encryption of these messages could be provided by using 3DES encryption.

Since the notification is electronically signed and its authenticity and integrity is automatically checked, it does require as little time of the mobile subscriber as possible. Therefore, we can state that Requirement I has been met. Because most current mobile phones support SIM AT, we can also conclude that Requirement II has been fulfilled. By checking the validity of the electronically signed notification message the SIM AT application is able to check the authenticity (only the service provider can make a valid signature) and the integrity of the notification message automatically. Therefore, we conclude that Requirement III has been fulfilled.

Our solution has several shortcomings regarding the availability of the notification service. The used SMS service does not provide acknowledgements for delivered messages and is dependent on the availability of the infrastructure of the mobile operator. If the mobile subscribers phone is unreachable or even switched off, a notification in time is impossible. Furthermore, SMS depends on unused capacity in the signaling channels and has consequently low priority within the channel utilization [31]. However, additional steps could be undertaken to improve the probability that the mobile subscriber receives the needed information in time. For example, different channels could be used, like sending the notification via CBS, SMS as well as e-mail. Furthermore, the availability of any service during disasters is dependent on the robustness of the underlying infrastructure [24]. Many disasters have had direct impact on communication infrastructures within disaster areas. In centralized communication infrastructures, disasters also have effects on the non-directly effected parts of the infrastructure if nodes are in the disaster area. One example is the failure of communication infrastructures during and after the hurricane "Katrina" in 2005 [20]. Network providers anticipated the probability of such an event in this area and protected their underlying communication infrastructure against outer influencing factors like heat, humidity, dust, and mud.

Additionally, mobile networks have been abused by terrorists, e.g. for setting off bombs like in Madrid in 2004 [20]. To prevent further explosions, officials claim the need to be able to shut down mobile networks in an emergency case. Schneier [32] argues that this would not significantly reduce the risk of further detonations. Terrorists might anticipate this behavior and might use alternative ways, like kitchen timers, for setting off bombs. He also concludes that victims benefit far more from telecommunication infrastructures than attackers do. Summarizing, we conclude that Requirement IV cannot completely be fulfilled. When communicating with specialists, confidentiality can be provided by ciphering the data using Wireless Transport Layer Security (WTLS), Secure Socket Layer (SSL). Therefore, Requirements V and VI have been fulfilled. By signing the messages to the disaster manager, the specialist ensures their integrity and authenticity. Therefore, we can state that Requirement VII has been met. The proposed solution has almost fulfilled the requirements that we defined in the previous section, except Requirement IV (Availability).

5 Conclusion

With this contribution, we proposed a DMS based on mobile communication infrastructures. In order to use such a system in the real world, security requirements such as availability, accountability, integrity and confidentiality have to be ensured by the DMS. Therefore, we derived security requirements for notifying civilians and specialists. Based on these requirements we proposed further refinements of the infrastructure, that address these issues. These refinements are based on SIM cards that are capable to create and verify electronic signatures. Such SIM cards are not widely deployed, but the technology exists. How to achieve a broad diffusion of such SIM cards is outside the scope of this paper.

However, not all requirements could reliably be achieved by using one single communication infrastructure but disaster management would clearly benefit from communication infrastructures that are more sophisticated.

References

1. 3G (2005) 1.5 Billion GSM Wireles Customers Across the Globe, *http://www.3g.co.uk/PR/Sept2005/1875.htm,* accessed 9 September 2005.
2. Abdulharis, R., Hakim, D., Riqqi, A. and Zlatanova, S. (2005) Geo-information as Disaster Management Tools in Aceh and Nias, Indonesia: a Post-disaster Area: Workshop on tools for emergency and disaster management, Brno.
3. Barry, J. M. (1998) Rising Tide: The Great Mississippi Flood of 1927 and How It Changed America, Simon & Schuster.
4. Borking, J. (1996) Der Identity Protector, *Datenschutz und Datensicherheit (DuD),* 20, 11, 654-658.
5. Burke, K. and Yasinsac, A. (2004) The ramifications of E911, College of Arts and Science, Tallahassee, Florida, USA.
6. Büllingen, F. and Stamm, P. (2004) Mobile Mulitmediadienste, Deutschlandschancen im internationalen Wettbewerb: Eine Internationale Vergleichsstudie, Bad Honnef.
7. Dietrich, V. J. (2004) Die Wiege der abendländischen Kultur und die minoische Katastrophe - ein Vulkan verändert die Welt, Alpnach Dorf.
8. Ellington, B. (2004) Enhancing E911 Systems a usability plan, *Proceedings of the Tenth Americas Conference on Information Systems (AMCIS 2004),,* August 2004, New York, ACM, 3419 - 3425.
9. European Commission (2003) COMMISSION RECOMMENDATION on the processing of caller location information in electronic communication networks for the purpose of location-enhanced emergency call services, Official Journal of the European Union, Brussels.
10. European IST Project 'Wireless Trust for Mobile Business' (WiTness) (2004) SIM Application Hosting - Detailed description of the concept, www.wirelesstrust.org/publicdocs/Witness_32275_D4_ExecSum.pdf, March 2005.
11. European Telecommunications Standards Institute (1992) GSM 3.40 - Technical Realization of the Short Message Service - Point-to-Point.
12. European Telecommunications Standards Institute (ETSI) (2003) Requirements for communication of citizens with authorities/organisations in case of distress (emergency call handling), ETSI SR 002 180 V1.1.1 (2003-12),ETSI, Sophie-Antipolis.
13. European Telecommunications Standards Institute (ETSI) (2004) Requirements for communication between authorities/organisations during emergencies, DRAFT ETSI SR 002 181 V0.3.0 (2004-12), ETSI, Sophie-Antipolis.
14. European Telecommunications Standards Institute (ETSI) (2004) Requirements for communications between citizens during emergencies, Draft ETSI SR 002 410 V0.0.1 (2004-09),ETSI, Sophie-Antipolis.
15. European Telecommunications Standards Institute (ETSI) (2004) Requirements for communications from authorities/organisations to the citizens during emergencies, DRAFT ETSI SR 002 182 V0.1.3 (2004-11),ETSI, Sophie-Antipolis.
16. EWCII (2003) Integrating Early Warning into Relevant Policies, Bonn

17. Forsyth, P. Y. (1999) Thera in the bronze age, Lang, New York u.a.
18. Fritsch, L. and Scherner, T. (2005) A Multilaterally Secure, Privacy-Friendly Location-based Service for Disaster Management and Civil Protection, *Proceedings of the AICED/ICN 2005, (LNCS 3421)*, Berlin, Heidelberg, New York, Springer, 1130-1137.
19. GSM Association (2006) GSM Coverage Maps, *http://www.gsmworld.com/roaming/gsminfo/index.shtml,* accessed 22.02.2006.
20. GSM World Series online "A Week in Wireless #194,"2005.
21. Ghosh, A. G. J. "A Strike At Europe's Heart, "TIME Europe2004, online version.
22. Ian Harris (2005) LS - Use of SMS and CBS for Emergencies: Technical Specification Group Terminals TSGT#27(05)0051, Meeting #27, 09 - 11 March 2005, ETSI, Tokyo.
23. Lane, N. (2000) Effective Disaster Warnings, Subcommittee on Natural Disaster Reduction, Working Group on Natural Disaster Information Systems, Washington.
24. Little, R. G. (2003) Toward More Robust Infrastructure: Observations on Improving the Resilience and Reliability of Critical Systems, in IEEEE (Eds.), *Proceedings of the 36th Hawaii International Conference on System Sciences,* Hawai, IEEE.
25. Ludden, B., Pickford, A., Medland, J. and Johnson, H.(2002) Cgalies final report V1.0, Report on implementation issues related to access to location information by emergency services (E112) in the European Union.
26. McNeil, D. G. J."What happens after disaster? Calamity, or reason to unite?" Milwaukee Journal Sentinel 2005.
27. Muntermann, J. and Rossnagel, H. (2006) Security Issues and Capabilities of Mobile Brokerage Services and Infrastructures, *Journal of Information System Security,* 2, 1.
28. Rannenberg, K. (2000) Multilateral Security - A concept and examples for balanced security, *Proceedings of the 9th ACM New Security Paradims Workshop,* Cork, Ireland, ACM Press, 151-162.
29. Rivest, R. L., Shamir, A. and Adleman, L. (1978) A Method for Obtaining Digital Signatures and Public Key Cryptosystems, *Communications of the ACM,* 21, 2, 120-126.
30. Scherner, T. and Fritsch, L. (2005) Notifying Civilians in Time - Disaster Warning Systems Based on a Multilaterally Secure, Economic, and Mobile Infrastructure, in AMCIS (Eds.), *Proceedings of the Eleventh Americas Conference on Information Systems (AMCIS) 2005.*
31. Schiller, J. (2003) Mobilkommunikation, Pearson Studium, München.
32. Schneier, B. (2005) Schneier on Security; A weblog covering security and secuirity technologies: Turning Cell Phones off in Tunnels, *http://www.schneier.com/blog/archives/2005/07/turning_cell_ph.html,* accessed July 19, 2005.
33. September 11th.com (2003) September 11, 2001-The Day the World Changed, *http://www.september11news.com/,* accessed 01 January 2006.
34. Spiegel online (2004) The Wall of Water: Part I, *http://service.spiegel.de/cache/international/spiegel/0,1518,335281,00.html,* accessed December 31, 2004.
35. Visscher, R. (2000) Chernobyl Nuclear Disaster, *http://www.chernobyl.co.uk/chernobyl_in_the_news.htm,* accessed 21st February 2006.
36. Yuan, Y. and Detlor, B. (2005) Intelligent Mobile Crisis Response Systems,: Systems to help coordinate responder communication and response in order to minimize the threat to human life and damage to property, *Communications of the ACM, Volume 48, Issue 2* February 2005, New York, ACM Press, 95- 98.
37. van der Togt, R., Beinat, E. Z. S. and Scholten, H. J. (2005) Location Interoperability Services for Medical Emergency Operations during Disasters, in P. van Oosterom, S. Zlatanova and E. M. Fendel (Eds.), *Geo-information for Disaster Management,* Heidelberg, Springer Verlag, 1127 - 1141.

A Fair Anonymous Submission and Review System

Vincent Naessens[1], Liesje Demuynck[2,*], and Bart De Decker[2]

[1] KULeuven Campus Kortrijk, Department of Computer Science,
E. Sabbelaan 53, 8500 Kortrijk, Belgium
`vincent.naessens@kuleuven-kortrijk.be`
[2] KULeuven, Department of Computer Science,
Celestijnenlaan 200A, 3000 Heverlee, Belgium
{`liesje.demuynck, bart.dedecker`}`@cs.kuleuven.be`

Abstract. Reputation systems play an important role in many Internet communities. They allow individuals to estimate other individual's behavior during interactions. However, a more privacy-friendly reputation system is desirable while maintaining its trustworthiness.

This paper presents a fair anonymous submission and review system. The review process is reputation-based and provides better anonymity properties than existing reputation systems. Moreover, the system allows for accountability measures. Anonymous credentials are used as basic blocks.

1 Introduction

In science, peer review is the oldest and best established method of assessing manuscripts, applications for research fellowships and research grants. However, the fairness of peer review, its reliability and whether it achieves its aim to select the best scientist or contributions has often been questioned. It is widely believed that *anonymous reviewing* helps fairness, by liberating reviewers from the fear that openly stated criticism might hurt their careers. Some researchers may be reluctant to write negative reviews as it could hamper future promotions.

Moreover, Bornmann et al. [1] argue that reviewer's recommendations are frequently biased, i.e. judgements are not solely based on scientific merit, but are also influenced by personal attributes of the author such as author's institution or name. *Anonymous submissions* can tackle this problem.

On the other hand, Meyer [5] suggests that referees too often hide behind anonymity to turn in *sloppy reviews*; worse, some dismiss contributions unfairly to protect their own competing ideas or products. Even people who are not fundamentally dishonest will produce reviews of unsatisfactory quality out of negligence, laziness or lack of time because they know they can't be challenged. Thus, referees/reviewers must be encouraged to do a decent job. If not, it must still be possible to hold them accountable.

---

* Research Assistant of the Research Foundation - Flanders (FWO - Vlaanderen).

H. Leitold and E. Markatos (Eds.): CMS 2006, LNCS 4237, pp. 43–53, 2006.

This paper presents a fair anonymous submission and review system. It achieves a reasonable trade-off between the anonymity requirements of the authors and reviewers and still allows to identify unfair reviewers. The proposed system aims at improving the fairness of review processes.

The rest of this paper is organized as follows: section 2 describes a general anonymous credential system; these credentials will be used in the submission/review system that is designed in section 3. Section 4 evaluates the system and points to related work. Section 5 concludes with a summary of major achievements.

2 Anonymous Credentials

Anonymous credentials allow for anonymous yet accountable transactions between users and organizations. In this section, a simplified version of the Idemix anonymous credential system [2, 7] is presented and extended with a new protocol for credential updating. The protocols are used as basic building blocks in our system. They typically run over an anonymous communication channel.

RegNym Protocols. An individual can establish multiple non-transferable pseudonyms (i.e. *nyms*) with the same organization. Two registration protocols are discussed:

- $U \leftrightarrow O$: $(Nym_{UO}, Sig_{UO}) = RegSignedNym(Cert_{UA})$. During the signed nym registration protocol, the user signs the established Nym_{UO} with his signature key, which is certified through an external certificate (which links the user's public key with his identity). Hence, the organization holds a provable link between the nym and the identity certified by the certificate.
- $U \leftrightarrow O$: $Nym_{UO} = RegNym()$. The (ordinary) Nym Registration protocol is used to register a regular nym between a user U and an organization O.

ProofNymPossession Protocol. $U \leftrightarrow O : ProofNymPossession(Nym_{UO})$. A user U can prove to an organization O to be the owner of a nym Nym_{UO}.

Issue Protocol. $U \leftrightarrow I$: $Cred_{UI} = IssueCred(Nym_{UI}, sl, \{attrName = attr\text{-}Value, \dots\})$. An issuer I can issue a credential $Cred_{UI}$ to a nym Nym_{UI}. The retrieved credential is known only to the user and cannot be shared. During the issue protocol, the showlimit sl of the credential is set to be either a constant k or unlimited. Also, a number of attributes is embedded into the credential.

Show Protocol. $U \leftrightarrow V$: $Transcript_{UV} = ShowCred(Cred_{UI}, [Nym_{UV}], [Dean\text{-}Cond], [AttrProperties], [Msg])$. A user U proves to a verifier V that he is in possession of a valid credential $Cred_{UI}$. This action results in a transcript for the verifier. During the protocol, several options may be enabled. The user may show his credential with respect to a pseudonym Nym_{UV}, by which he is known to V. This provably links the transcript and the nym. In addition, the resulting transcript may be deanonymizable: upon fulfillment of a condition *DeanCond*, a trusted deanonymizer is allowed to recover the nym on which the credential was issued. Moreover, the user may disclose some information about the attributes encoded into the credential. He may reveal either an attribute or a property of the attribute, and may decide to sign a message *Msg* with his credential; creating a provable link between the transcript and the message. Note that different transcripts for the same credential cannot be linked (unless the value of a unique attribute is proved), nor can they be linked to the credential's issue protocol.

Update Protocol. A user U can update his credential $Cred_{UI}$ by interacting with its original issuer I. This is particularly useful when the credential has attributes of which the value may change over time. The protocol consists of the user showing his credential to I and consecutively receiving a new credential (i.e. the actual update). The new credential is issued on the same nym as the old credential. Its attributes are either the attributes of the old credential or the result of a simple operation f on these attributes (e.g, adding a known value). Apart from the public parameters of the operation f and what is explicitly revealed by the user, the issuer does not have any information about the new credential's attributes. Note that the old credential will still be valid after the execution of the protocol unless it is a one-show credential. Note also that an *UpdateCred* protocol can never be executed without a preceding *ShowCred* protocol.
$U \leftrightarrow I$: *Transcript*$_{UI}$ = *ShowCred(Cred$_{UI}$,Nym$_{UI}$, [DCond], [AttrProps], [Msg])*
$U \leftrightarrow I$: *UpdateCred(Cred$_{UI}$,sl,[AttrChanges])*

Local Deanonymization Protocol. *D: (Nym$_{UI}$, DeAnProof) = DeanonLocal(Transcript$_{UV}$).* If a credential show is deanonymizable, the pseudonym *Nym$_{UI}$* on which the credential was issued can be revealed by a trusted deanonymizer D. *DeAnProof* proofs the link between the transcript and the nym. D is only allowed to perform the deanonymization when *DeanCond* is met.

3   A Fair Anonymous Submission and Review System

First, the requirements and roles are described. Next, we describe the protocols used in the different phases. Finally, complaint handling procedures are discussed.

3.1   Requirements and Roles

Requirements. Whereas current conference systems mainly focus on the anonymity requirements of the authors, our design considers the concerns of all users:

- *Anonymity requirements.* Committee members (i.e. reviewers) must be able to review papers anonymously. Similarly, authors must be able to submit papers anonymously. The identity of authors may only be disclosed when the paper is accepted (i.e. the identity of the authors is required for preparing the program) or when the paper is submitted simultaneously to another conference (i.e. no conference chair accepts double submissions).
- *Requirements related to fairness.* First, committee members are not allowed to review the same paper multiple times or to advice on their own papers.

Second, the identity of a reviewer can be disclosed if he has written many *unacceptable* reviews. Third, the reviewers' familiarity with the research domain must have an impact on the final outcome of the review process. Therefore, reviewers may not be able to lie about their expertise. Finally, committee members must be encouraged to review the papers that are assigned to them. For instance, they can get a discount on the conference fee.

Roles. *Users (U)* are either authors or reviewers. The *Reputation manager (R)* initializes and updates their reputations. The reputation manager is independent of any conference system.

The *Conference system* is administered by the Conference Chairman *(C)*. As depicted in figure 1, the conference system consists of a front end and a back end. The *front end* of the system consists of three parts: a submission manager, a review manager and a complaint manager. The *submission manager* handles requests from authors. Authors can submit papers and retrieve a contribution token when their paper is accepted. The *review manager* handles requests from reviewers. Reviewers can register as a committee member. Thereafter, they can review papers. Finally, they can retrieve a discount token. The *complaint manager* handles complaints from both authors and reviewers. The *back end* of the conference system consists of a storage manager. The storage manager is responsible for storing submitted papers, reviews and certain types of evidence.

There is also a *deanonymization infrastructure*. It consists of an *Arbiter (A)* and a *Deanonymizer (D)*. *A*'s role is to verify whether a de-anonymization condition is fulfilled. *D* can retrieve the pseudonym under which a credential is issued from a "show"-transcript. An *anonymous communication infrastructure (=AC)* is required as the connection between *U* and the conference system needs to be anonymous.

3.2   Protocols

This section describes the protocols used in different phases. The relation between the protocols are shown in figure 2.

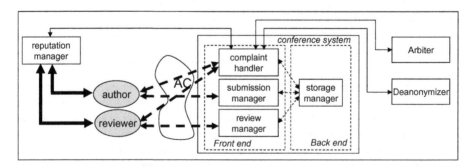

Fig. 1. Overview of the Conference Management System

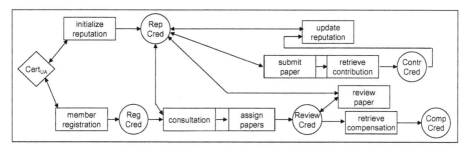

Fig. 2. Overview of actions and credential types

Initialize_reputation. In this phase, a new researcher (i.e. U) contacts the Reputation Manager R to initialize his reputation in a field. The user first establishes a nym and signs that nym with an external certificate (issued by a trusted certificate authority A). R stores the identity proof and issues a reputation credential on the nym. It can be shown unlimitedly. Note that an individual can retrieve new reputation credentials as he explores new research domains.

$U \leftrightarrow R : (Nym_{UR}, Sig_{UR}) = RegSignedNym(Cert_{UA})$
$U \leftrightarrow R : RepCred_{UR} = IssueCred(Nym_{UR}, *, \{repField = field, repValue = 0\})$
R: stores $\{Nym_{UR}, Sig_{UR}, Cert_{UA}\}$

Submit_paper(paper). Submitting a paper is conditionally anonymous. When an author submits a *new* paper[1], he remains anonymous as long as his paper is not accepted. The author first establishes a nym with the submission manager. During the credential show, the paper is signed, which provably links the paper to the transcript of the credential show. The transcript is deanonymizable. The Submission Manager verifies the credential show and passes it to the Storage Manager. The Storage Manager stores the paper, the nym and the transcript.

$U \leftrightarrow C : Nym_{UC} = RegNym()$
$U \leftrightarrow C : Transcript_{UC} = ShowCred(RepCred_{UR}, Nym_{UC},$
    $'accepted \vee double\_submission', null, \{paper\})$
C: stores $\{Nym_{UC}, Transcript_{UC}, paper\}$

Retrieve_contribution. After the review process, the Conference Manager publishes a whitelist containing the titles of accepted papers. Each title contains a link to a nym Nym_{UC} used during submission. The author can check whether his paper is accepted. If so, he contacts the Submission Manager, proves his identity and also proves to be the owner of the corresponding Nym_{UC}. The Submission Manager verifies the proof and issues a contribution token $ContrCred$ to the

---

[1] A *new* paper is a paper that is not sent previously to another conference. When the paper has already been submitted to another conference, and this is detected, the author's identity will be revealed.

author. The author can use this one-show credential once, to update his reputation. It contains two attributes: the conference id and the research field of the accepted paper.

Note that an author who forgets to check the whitelist can still be traced. A deanonymizer will eventually reveal the identity of an author. Hereto, the conference manager must convince the deanonymizer that the paper is really accepted. The strategy to reveal the author behind a submission is discussed in section 3.3.

$$U \leftrightarrow C : Sig'_{UR} = ProofIdentity(Cert_{UA})$$
$$U \leftrightarrow C : ProofNymPossession(Nym_{UC})$$
$$U \leftrightarrow C : ContrCred_{UC} = IssueCred(Nym_{UC}, 1,$$
$$\{contrConf = conf, \; contrField = field\})$$

Update_reputation(field,delta). The researcher presents a reputation credential and a contribution credential. Moreover, he proves that the research field in the reputation credential corresponds to the research field in the contribution credential. The reputation credential is updated with a *delta* value. The *delta* value can depend, among others, on an international ranking.

$$U \leftrightarrow R : Transcript_{UR} = ShowCred(RepCred_{UR}, \; null, \; null,$$
$$\{repField == field\}, \; null)$$
$$U \leftrightarrow R : Transcript'_{UR} = ShowCred(ContrCred_{UC}, \; null, \; null,$$
$$\{f(contrConf) == delta \wedge contrField == field\}, \; null)$$
$$U \leftrightarrow R : UpdateCred(RepCred_{UR}, \; ^{*}, \; [repValue+=delta])$$

Member_registration. In this step, each committee member (i.e. U) contacts the Review Manager to retrieve a registration credential. The committee member first establishes a Nym'_{UC} and signs that nym with an external certificate. The registration credential will be used to control the consultation process.

$$U \leftrightarrow C : (Nym'_{UC}, \; Sig'_{UC}) = RegSignedNym(Cert_{UA})$$
$$U \leftrightarrow C : RegCred_{UC} = IssueCred(Nym'_{UC}, 1, \{revConf = conf\})$$
$$C : \text{stores } \{Nym'_{UC}, Sig'_{UC}, Cert_{UA}\}$$

Consultation(preferences). After the registration deadline, registered members can specify their individual *preferences* anonymously (i.e. relative to a Nym''_{UC}). Reviewers are required to prove that they have enough experience (i.e. a high reputation) in that field, before they are allowed to bid on a paper. The Review Manager processes the preferences of all committee members.

$$U \leftrightarrow C : Nym''_{UC} = RegNym()$$
$$U \leftrightarrow C : Transcript^1_{UC} = ShowCred(RegCred_{UC}, \; Nym''_{UC}, \; \text{'multiple-}$$
$$unacceptable\_reviews', \{revConf == conf\}, \; preferences)$$
$$U \leftrightarrow C : Transcript^2_{UC} = ShowCred(RepCred_{UR}, \; Nym''_{UC}, \; null,$$
$$\{repField == field \wedge repValue > x\}, null)$$
$$C : \text{stores } \{Nym''_{UC}, \; Transcript^1_{UC}, \; preferences\}$$

Assign_papers. In this step, each committee member (i.e. U) contacts the Review Manager to retrieve a review credential. The committee member proves to be the owner of a Nym''_{UC} (that was established in the consultation phase). The review credential will be used to control the review process. A review credential contains a set of paper identifiers *revS*. The committee member is expected to review each of the papers that correspond to the identifiers. It is clear that *revS* depends on the preferences of the reviewer.

$U \leftrightarrow C : ProofNymPossession(Nym''_{UC})$
$U \leftrightarrow C : ReviewCred_{UC} = IssueCred(Nym''_{UC},1, \{revConf = conf, revS = S\})$

Review_paper(paperId). The reviewer submits his advice on a paper during this phase. An advice typically consists of a list of comments and a score on multiple evaluation criteria (originality, readability...).

The Committee Member shows his review credential to prove that the *paper* for which he wants to submit an *advice* was assigned to him. The reviewer can also choose to prove that his reputation is higher than some predefined level. This allows the Conference Chairman to measure the familiarity of the reviewer with the research domain of the paper. If the advice is submitted successfully, the Review Manager updates the members' review credential (i.e. the *paperId* is removed from the list of assigned papers). As each review credential is a one-show credential, the old review credential becomes useless. Therefore, a reviewer cannot comment multiple times on the same paper.

$U \leftrightarrow C : Transcript^3_{UC} = ShowCred(ReviewCred_{UC},\ null,$
$\qquad\qquad 'unacceptable\_review',\ \{paperId \in revS \land revConf == conf\},\ null)$
$U \leftrightarrow C : Transcript^4_{UC} = ShowCred(RepCred_{UR},\ null,\ null,$
$\qquad\qquad \{repField == field \land repValue>x\},\ \{advice,\ paperId\})$
$U \leftrightarrow C : UpdateCred(ReviewCred_{UC},\ 1,\ \{revS = revS \setminus paperId\}\ )$
$C : stores\ \{advice,\ paperId,\ Transcript^3_{UC}\}$

Retrieve_compensation. After the review deadline, the committee member finalizes his job by contacting the Review Manager. The reviewer first proves to be the owner of a Nym'_{UC}. As explained before, Nym'_{UC} can be provably bound to the identity of a Committee Member. Next, he submits his review credential to prove that the set of remaining papers *revS* is empty. Hence, the Conference Chair knows which committee members have finalized all reviews. This allows the Conference Manager to send a reminder to members that haven't finalized the reviews if the deadline is passed. Optionally, the Review Manager issues a compensation token that can be used to get a discount on the conference fee.

$U \leftrightarrow C : ProofNymPossession(Nym'_{UC})$
$U \leftrightarrow C : Transcript^5_{UC} = ShowCred(ReviewCred_{UC},\ Nym'_{UC},$
$\qquad\qquad null,\{revS == \emptyset\},\ null)$
$U \leftrightarrow C : CompCred_{UC} = IssueCred(Nym'_{UC},\ 1,\ null)$

3.3   Complaint Handling

Two types of complaints are discussed in this section: complaints related to submissions and complaints related to reviews. A *submission is unacceptable* if it is sent previously/simultaniously to another conference. If so the identity of the author must be revealed[2]. It consists of three steps:

- **Decision of Arbiter (A).** The Complaint Handler sends the suspected paper(s) to A. A verifies whether the papers are really very similar and returns his signed decision. If so, the Complaint Handler informs D.
- **Disclosing Nym.** D receives a signed message from the Complaint Handler. The message contains A's decision, the paper and the $Transcript_{UC}$. D verifies the decision, and if positive, deanonymizes the transcript. He then returns Nym_{UR} and a deanonymization transcript to the Complaint Handler.
- **Revealing identity.** The Complaint Handler forwards the evidence to the Reputation Manager R and orders R to reveal the identity of the user behind the Nym_{UR}. The Complaint Handler stores the evidence that proves the link between the author and the submissions.

An *unacceptable review policy* can be worked out by the Conference Manager. Note that both a conference chairman as well as an author (when receiving feedback) can initiate a complaint of this type. It consists of three steps:

- **Decision of Arbiter.** (see above[3])
- **Disclosing review identifier.** If the review is unacceptable, the Complaint Handler convinces D to deanonymize $Transcript^3_{UR}$. D then returns the Nym''_{UC} and the deanonymization transcript.
- **Revealing identity (optionally)** If multiple unacceptable reviews correspond to the same Nym''_{UC}, the Complaint Handler sends the evidence and $Transcript'_{UC}$ to D. D deanonymizes $Transcript'_{UC}$ and returns Nym'_{UC} and the deanonymization transcript to the Complaint Manager. The Conference system keeps a provable mapping between Nym'_{UC} and the identity of the reviewer.

4   Evaluation

This section focuses on the anonymity/trust properties of the system. The conference management system creates a trusted environment for all players.

An *author* may trust that his submission will not be linked to his identity (even not by the conference chairman) as long as his paper is not accepted and not double submitted. Four entities are required to reveal the identity of an author, namely C, A, D and R. D will only deanonymize the transcript after permission of an arbiter. However, trust can easily be distributed between multiple deanonymizers D_i and arbiters A_j. This implies that a set of arbiters decide whether the deanonymization condition is fulfilled and a set of deanonymizers is required to reveal the nym_{UR} behind the transcript.

---

[2] Note that this strategy can also be used to reveal the author of an accepted paper who forgot to check the whitelist.

[3] Note that in this case, the suspected review is sent to the Arbiter.

Except in very unusual circumstances, the identity of the *reviewers* involved in the review of any given paper is not known by any party. The identity of reviewers will only be revealed if they wrote several reviews of inferior quality. C, A and D are required to disclose the identity of a reviewer. Again, trust can be distributed between multiple arbiters and deanonymizers.

Although the *conference manager* does not know the identity of the reviewer of a paper, a referee can not lie about his expertise in a research domain. This improves the fairness of the review proces.

Researchers can only update their reputation if they retrieved a contribution credential. Consequently, the *reputation manager* needs to rely on conference managers. However, the reputation manager will only increase the users reputation value slightly if the contribution credential was issued by a low ranked conference.

5   Discussion

Anonymous reputation systems [3, 4, 6] already play an important role in Internet communities like eBay. Unfortunately, the design of current reputation systems allows to generate user profiles. Ultimately, the user can be uniquely identified. The main problem is that the reputation is tightly-coupled to a pseudonym in many systems. Our design does not bind a reputation value to a single pseudonym. Thus, multiple proofs of the same reputation cannot be linked. Moreover, our system enables to prove properties of the reputation value (for instance, *value > 10*). This implies that a user with a very high reputation value can still convince a conference chairman without being uniquely identified. We have demonstrated the use of updatable credentials within an anonymous reputation system. It is clear that this new concept is useful in many applications. In particular in applications where the value of a credential's attribute depends on external factors and hence may change over time. In its low-level implementation, a show protocol precedes the actual update protocol. Its computational cost is slightly more than the cost of an individual show or issue protocol, but significantly less than the cost of both primitives together.

The *reputation credential* and the *review credential* contain attributes whose value can change. However, both types have a slightly different implementation. Whereas reputation credentials are multi-show credentials, review credentials are one-show. Both strategies have advantages that are exploited in the conference system. An unlimited-show credential allows users to prove (properties of) attributes unlimitedly. Hence, researchers can prove properties of their expertise without having their credential to be updated. One-show credentials prevent subjects to use the credential multiple times. Thus, a committee member cannot present an older version of the review credential multiple times (i.e. $ReviewCred_{UC}$). This prevents him to submit more than one review for the same paper. However, a reviewer can use an old reputation credential (i.e. $RepCred_{UR}$)

when reviewing a paper. Nevertheless, as newer reputation credentials have a higher value, he will not be inclined present an older one.

Researchers can have *expertise in multiple research domains*. Similarly, a paper can present experiences in multiple research domains. Hence, a reviewer must be able to prove his familiarity with each of these domains. In a straightforward solution, the researcher retrieves a reputation credential from the Reputation Manager for each domain in which he is involved and uses a subset of these credentials at each review. This has many disadvantages. First, a mature researcher may have to store many credentials. Second, a lot of overhead is introduced when multiple reputation credentials have to be shown. Another solution foresees multiple domains and values in one credential. However, as many research domains exist, the credential size will also be large. A hybrid solution defines a set of general research domains. Each domain is split in subdomains. A credential can be retrieved for each domain. One credential stores a researchers' reputation value within each subdomain. For instance, the ACM Computing Classification System can be used to fix sub(domains).

If an individual has not made any relevant contributions within the last years, his *reputation value may be misleading*. This can compromise the fairness of the review process. To tackle this problem, the reputation credential could also keep the dates and contribution values of the most recent publications. These attributes can also be used to calculate the user's final reputation value. Hence, reputation credentials that are not updated recently decrease implicitly: $f(value, [year_1, value_1], [year_2, value_2], [year_3, value_3]) > x$.

Although a conference manager can demand from committee members to indicate *conflicting interests* during the consultation phase, a committee member can still neglect this demand. Hence, a committee member could be assigned his own paper. However, the authors behind accepted papers are identified. Moreover, the Conference Chairman stores the nyms Nym''_{UC} of reviewers that did comment on a paper. He also stores the corresponding $Transcript^1_{UC}$ which can be deanonymized by D (and which can lead to the identity of the reviewer). Consequently, D can check after the review process whether a conflict of interests occurred. If so, he informs the Conference Chairman who, on his turn, can decide to revise the acceptability status of the paper. Alternatively, authors can be demanded to indicate conflicts of interests. However, the latter strategy may reduce the anonymity set of authors.

6   Conclusions

This paper presented a fair anonymous submission and review system. The system provides a trusted environment for authors, reviewers and conference chairmen. The review process is reputation-based and allows for accountability measures. We also demonstrated the use of updatable credentials within an anonymous reputation system. It is clear that this new concept can be extended to many other application domains where the value of a credential's attribute depends on external factors and hence may change over time.

References

1. L. Bornmann, H. D. Daniel, Reliability, fairness and predictive validity of committee peer review. Evaluation of the selection of post-graduate fellowship holders by the Boehringer Ingelheim Fonds.B.I.F. In *FUTURA 19*, p. 7-19, 2004.
2. Jan Camenisch, Els Van Herreweghen. Design and Implementation of the Idemix Anonymous Credential System. Research Report RZ 3419, IBM Research Division, June 2002. Also appeared in *ACM Computer and Communication Security*, 2002.
3. C. Dellarocas. Immunizing online reputation reporting systems against unfair ratings and discriminatory behavior. In *Proceedings of the ACM Conference on Electronic Commerce*, 2000, 150-157.
4. R. Dingledine, N. Mathewson, and P. Syverson. Reputation in P2P Anonymity Systems. In *Proc. of Workshop on Economics of Peer-to-Peer Systems*, June 2003.
5. B. Meyer. Open refereeing. http://se.ethz.ch/~meyer/publications/
6. S. Steinbrecher. Privacy-enhancing Reputation Systems for Internet Communities. In *Proceedings of the 21th IFIP International Conference on Information Processing: Security and privacy in dynamic environments*, to appear, May 2006.
7. E. Van Herreweghen. Unidentifiability and Accountability in Electronic Transactions. PhD Thesis, KULeuven, October 2004.

Attribute Delegation Based on Ontologies and Context Information

Isaac Agudo, Javier Lopez, and Jose A. Montenegro

Computer Science Department, E.T.S. Ingenieria Informatica
University of Malaga, Spain
{isaac, jlm, monte}@lcc.uma.es

Abstract. This paper presents a model for delegation based on partial orders, proposing the subclass relation in OWL as a way to represent the partial orders. Delegation and authorization decisions are made based on the context. In order to interact with the context, we define the Type of a credential as a way to introduce extra information regarding context constraints. When reasoning about delegation and authorization relationships, our model benefits from partial orders, defining them over entities, attributes and the credential type. Using these partial orders, the number of credentials required is reduced. It also classifies the possible criteria for making authorization decisions based on the context, in relation to the necessary information.

1 Introduction

This work presents a delegation model that defines general credentials. A credential is defined as a link between two entities, in relation with some attribute. Credentials also have a type, which defines their special characteristics and in particular, information regarding the context. Two of these characteristics are whether it is a delegation or an authorization credential and the validity time interval.

We use the word Delegation to describe the transfer of management rights over certain attributes. The sentence "A delegates attribute r to B" is used as a shortcut for "A authorizes B to issue credentials regarding the attribute r to any other entity C". We use delegation as a Meta concept in an authorization framework, because delegation statements are authorization statements over the act of authorizing. This meta information is used to facilitate the distribution of authorization, providing a powerful instrument to build distributed authorization frameworks.

In real organizations, there are several variables that need to be considered when taking authorization decisions. It is important to define the different kinds of entities in the system and the relationships between them. We need and Organization Chart to be able to classify entities. This chart establishes a hierarchy of entities and all the decisions made within the organization have to comply with this hierarchy. There are cases in which this organization chart only has a few classes of entities and others in which there are a considerable number of

H. Leitold and E. Markatos (Eds.): CMS 2006, LNCS 4237, pp. 54–66, 2006.

them, but in any case the chart is quite static, so it is not worth using certificates or credentials to define it. One possible solution is to define an ontology of entities, represented as classes, using the subclass relation. This can be de done very easily using OWL [6] and any of the tools that support it. There is both an OWL plug in for Protege [5] and a stand alone application called SWOOP [2] that allows us to use OWL graphically.

Classes that are higher up in the hierarchy refer to more general classes and, on the other hand, those classes that are lower down in the hierarchy refer to more specific classes. So, if class A is a specialization of class B, then all privileges linked with B have to be inherited by A. Membership of entities is modeled using unitary classes, so we only talk about classes and subclass relations. This simple ontology simplifies the process of issuing credentials as we can issue privileges to several entities using only one credential and the OWL ontology. This ontology can be mapped to a partial order set in which the elements correspond to the OWL classes (unitary or not) and the subclass relation defines the partial order. We take the same approach with privileges. Attributes are used as a bridge to cover the gap between entities and privileges. We define a partial order over attributes as a hierarchy in which we go from general attributes (those more related to the concept of Role) to specific attributes (those more related to privileges and resources). With these ontologies, we simplify the delegation and authorization chart of the Organization, as we split it into three sections:

- Organizational relation between entities
- Organizational relation between attributes
- Specific delegation and authorization credentials.

The point here is to combine this information to make correct delegation and authorization decisions. Another interesting point of our proposal is the concept of Type, which is closely related to the context. In the Type of a credential we encode extra information that might affect delegation and authorization decisions depending on the context in which the decisions have to be made. The validity interval is an example of information that has to be included in the specification of the Type of a credential. We follow the same principle of using ontologies, in the definition of Types. In the simple example of time intervals, it is clear that we may establish a subclass relation over them, using the subset relationship. In general, to be coherent, all the information included in the Type is interpreted as a concept in a subclass only ontology. This ontology is used to automatically derive virtual credentials from real ones. Going back to the time intervals, if we chain two delegation credentials C_1 and C_2, which are valid in intervals t_1 and t_2 respectively, the validity interval of the path will be $t_1 \cap t_2$ as it is the maximal element that is both lower than t_1 and t_2. What we do here is to derive new credentials C_1 and C_2, both valid in $t_1 \cap t_2$, using the ontology (or partial order) defined for the type of credentials and in particular for validity intervals.

Therefore, in this paper we present a general framework to model both delegation and authorization relationships. This model is based on the definition of typed credentials, which is one containing extra information called the Type of the credential. This information is used to link the credential with its context.

We propose the definition of subclass relations over all the concepts used in the framework: Entities, Attributes and Types. These relations are encoded using ontologies, and in particular the OWL `subclassOf` relation. These subclass relations can be interpreted as a partial order relation[1] and in this paper we use only the mathematical notation, because it is easier to understand and work with. These ontologies provide helpful information for defining access control polices, simplifying the process of defining access credentials, control policies and making authorization and delegation decisions more efficient.

The definition of a credential type is particularly useful as it allows us to restrict delegation according to context parameters, such as location and time, and in general according to the state of the system. When working with delegation credentials, it very important to define the requirements that must hold for a credential to be valid, because revoking delegation credentials is very expensive task in terms of computation.

The structure of the paper is as follows. In section 2, we present the basic elements or building blocks of the framework: Entities, Attributes and Types, and explain the meaning of the ontologies (partial orders) defined for each of them. In section 3, credentials are defined as a construct that describe the basic elements of the framework. Credentials are defined as tuples of the form: $(Issuer, Holder, Attribute, Type)$. In this section we introduce the validity evaluation as a function that distinguishes between valid (active) and invalid (inactive) credentials in a given state and classify them according to the information used to make this decision. Section 4 defines paths or chains of delegation credentials, and explains how validity evaluation can be extended to define an evaluation function over credential paths. In section 5 some conclusions are presented.

2   Elements of the Framework

We mentioned before that a Delegation is a Meta-Authorization. So, before the definition of delegation credentials, we have to review authorization credentials.

The standard X.509 defines an authorization credential having the following fields[2]:

- *Issuer*: The entity who authorizes the Holder.
- *Holder*: The entity who receives the attributes.
- *Attributes*: The attributes issued to the Holder.
- *Validity Period*: The validity period of the authorization.

For a credential to be valid, the Issuer should have management rights over the attributes issued in the credential. In this case, the Holder is able to use the new attributes but it is not able to issue new credentials regarding these

---

[1] $A \leq B$ if and only if A is a subclass of B.
[2] We focus on the most important ones.

attributes. In the case where it is a Delegation credential, the Holder should be able to issue new (Authorization) credentials regarding these attributes, instead of being able to use the attributes directly. This is the main difference between an Authorization Credential and a Delegation Credential. Based on this definition, we define credentials as 4-tuples, in which we have an *Issuer* field, a *Holder* field, an *Attribute* field and a *Type* field.

2.1 Entities

Delegation is established between two entities. Thus, we have to define what an entity is in our framework. We distinguish between individual entities, those that are indivisible, and general entities (or just entities) that may be composed of more than one individual entity or refer to more complex concepts. We use this distinction inspired by the concept of Role [3]. In this sense, Roles are related to general entities. We define both general and individual entities as OWL classes. For example, in our framework Bob will be represented as a class, and the members of this class will be the instances of Bob, each time he is authenticated in the system.

In any working environment, there are some fixed rules that define entity relationships. If we think of a University, there is an inclusion relation between professors and employees, as professor is a specification of the concept (class) employee. So the classes *Professor* and *Employee* are related by the subClassOf relation. All privileges granted to employees should also be granted to professor. A shortcut to avoid this double issuing of credentials could be to define a partial order on entities. In this case, employee is a general entity but professor is an individual entity as there are no other entities more specific than professor. If we include real people in this simple example, the following chain of implications is obtained:

$$Alice \Rightarrow Profesor \Rightarrow Employee$$

Therefore neither Employee, nor Professor are individual entities, and only Alice has this category of individual entity.

Instead of using the symbol $\Rightarrow$ we use the symbol $\leq$ to emphasize that we have a partial order defined in the set of entities, we also use $<$ as a shortcut for $\leq$ and $\neq$. In this way, the previous chain of implications can be translated to:

$$Alice \leq Profesor \leq Employee$$

Definition 1 (Entities). *The set of entities is defined as a partial order* $(\mathcal{E}, \leq_e)$ *and the subset of individual entities by* $\mathcal{E}^* := \{e \in \mathcal{E} : \nexists e' \in \mathcal{E}, e' < e\}.$

Now, let us do it the other way round, and think about the following scenario. Suppose Bob needs a credential issued by an employee of the University to be able to enter the Library. If Alice issues this credential, the verifier of the credential, i.e. the library, should be able to compare *Alice* and *Employee* to determine if this credential is valid for this purpose. In this case, as $Alice \leq Employee$ a credential issued by *Alice* could also be interpreted as a credential issued by *Employee*.

So, if we have two entities E_1 and E_2 where $E_1 \leq E_2$ we obtain the following inference rules:

1. Any attribute issued to E_2 should also be issued to or used by E_1.
2. Any credential issued by E_1 could also be interpreted as a credential 'issued' by E_2.

Although only individual entities (minimal elements of $\mathcal{E}$) are allowed to issue credentials, we can give meaning to a credential 'issued' by non-individual entities. So a credential 'issued' by non-individual entities could be a requirement constraint for the authorization of some operations. This constraint means that an individual entity, which is lower in the hierarchy than the required non-individual entity, has to issue a credential in order to authorize the requested operation.

Depending on how many entities are ordered we get two extreme cases: a non ordered set and a complete lattice.

- Trivial poset. In this case, the set of entities in not really ordered. Thus, any entity is minimal and therefore all entities are individual entities, i.e $\mathcal{E}^* = \mathcal{E}$. This is the simplest approach and is the one used for example in SPKI [4].
- Power Set with the inclusion order. In this case $\mathcal{E} \simeq \mathcal{P}(\mathcal{E}^*)$. This structure allows us to use groups of entities as holders of credentials. The partial order is induced by the inclusion operator. In fact, the defined structure is a complete lattice, using the Union and Intersection operators as the join and meet operators. This case is not realistic due the huge number of classes that have to be considered.

Depending on the application context, the number of classes will vary. We have to reach a balance between the two extreme cases previously presented.

2.2   Attributes

In many authorization management systems, credentials grant access directly to resources, so there are no attributes. An attribute is a more general concept than resource. Resources can be easily identified in the system, but they do not provide us with a fine grain way of defining authorization. A Unix file, f, is clearly a resource, so an entity A could issue an authorization credential to B, regarding the resource f. In this case, shall B be able to modify it? In general it depends on the file. So, in many cases, resources are associated with different access modes. In order to describe an operation that could be authorized or not, we need a tuple $(resource, right)$. But resources are normally composed of lower level resources, e.g. a Computer is made up of memory, hard disk (which is also composed of files) and many other components. There is an implicit order relationship over the resources of a given system, as there is a partial order over entities. Because of the complexity of defining what is a resource and what not, it is better to use *privileges* or more accurately, *attributes* to define the nature of the authorization.

The use of attributes is an initiative to generalize the use of a tuple of the form $(resource, operation)$. Using attributes we give the authorization policy more relevance, as it is in the policy where we have to match attributes with the traditional tuple $(resource, operation)$. So attributes allow us to divide the authorization into two phases:

- Defining Attributes. It consists of giving a meaning to the attributes used in the system and therefore, it assigns a tuple $(resource, right)$ or a set of privileges to each attribute.
- Issuing Attributes. It consists of issuing credentials in relation with the previously defined attributes.

We identify two kinds of attributes widely used in the literature:

- Membership Attribute. It encodes the concept of *x is a member of role A*. Here the attribute is totally equivalent to the concept of Role.
- Specific Attribute. It encodes the concept of *x has access privilege A*. Here the attribute is totally equivalent to the privilege A which directly translates to a tuple of the form $(resource, right)$.

These two examples show that by using attributes we could model any of the existing traditional approaches. In our framework we can encode roles and role membership using entities and the partial order over entities, so we think of attributes as an intermediate concept between roles and privileges.

There are many situations in which there is a hierarchy over attributes. For example, write access to a file may imply read access. Another clear example is the case of attributes with parameters, if $AGE(x)$ represents that the owner entities are at least x years old, then $AGE(x)$ also implies $AGE(x-1)$ and in general $AGE(y)$ where $y \leq x$.

For this reason, we translate this hierarchy into a partial order over attributes using an OWL ontology similar to the one defined over entities. This ontology helps us in the decision making process.

If we think of a drug store, where alcohol is sold, prior to buying an alcoholic drink, entities should present a credential stating that they are at least 21 years old, i.e. $AGE(21)$. Now, suppose that an older person, with the attribute $AGE(60)$ tries to buy whisky. In our case, as his attribute is 'greater' than the required one,i.e. $AGE(21) \leq AGE(60)$, he should be allowed to buy whisky. Then, the requirement is the attribute $AGE(21)$ instead of the classical approach in which we require an attribute $AGE(x)$ where $x \geq 21$.

The privilege *BuyAlcohol* is defined as a specific attribute in the set of attributes $\{AGE(x) : x \in \mathbb{N}\} \bigcup \{BuyAlcohol\}$. The partial order is defined as follows: $BuyAlcohol \leq AGE(21)$ and $AGE(x) \leq AGE(y)$ if and only if $x \leq y$. This ontology helps us to understand the authorization policy in which only the attribute *BuyAlcohol* is required to buy alcohol.

2.3 Type

There are other parameters that have to be included in the specification of the credentials, besides the attribute, the issuer and the holder. We define the type of credential to include all the properties that are not essential, but which are helpful, to define Authorization and Delegation. This separation of concepts was previously proposed in [1], but we extend it here to include more general information, in particular, information regarding the context.

A credential type is mainly used to determine if it is valid or not under certain circumstances. Time is the basic type and it will be used as a model to define new types that help us to restrict delegation and authorization according to context information.

Consider the situation in which the only relevant characteristic to determine whether a credential is valid at a particular point in time is the validity interval. In this case, the type of the credential consists of a time interval. The set of time intervals has a natural partial order induced by the inclusion relation, i.e. one interval I_1 is lower than another interval I_2 if $I_1 \subset I_2$. Formally, $\mathcal{T}_I := \{[n, m] : n, m \in Time\_Instants \bigcup\{\infty\}, n \leq m\}$.

As with Entities and Attributes, this partial order can be used to derive new information from the current credentials of the system. Suppose we have a credential with type $[0, 5]$, then we could derive new credentials with a lower type, e.g. $[1, 4]$.

Another important type is the one that defines whether a credential is delegable or not. We define the type 0 for non delegable credentials and the type 1 for delegable credentials. Therefore, the delegation type is defined by $\mathcal{T}_D := \{0, 1\}$. If we prefer that delegation credentials explicitly imply authorization credentials, i.e. non delegable credentials, then we should define the partial order $0 \leq 1$, but in general we consider $\mathcal{T}_D$ not to be ordered.

We can now combine these two types and define a new type, $\mathcal{T}_{I \times D} := \mathcal{T}_I \times \mathcal{T}_D$, which includes both types with the natural partial order. We will describe more types in the following sections.

3   Credentials

At first sight, the information needed for a Delegation credential is the same as the information used to define an Authorization credential plus the delegation type that states whether it is a delegation or an authorization credential.In this way, we include the two types of credentials into one single concept. So hereinafter, credential is used as a general concept that comprises both authorization and delegation credentials.

If we look at the differences between delegation and authorization credentials, we see that the revocation of delegation credentials is more problematic than the revocation of authorization credentials. If we think of revocation of authorization credentials, we know this is a bottleneck for authorization frameworks. Using delegation credentials, a new problem arises however because when they are revoked, all the authorization and delegation credentials which are linked with it have to be revoked too. This is the chain effect.

In this situation, we need some mechanisms in order to minimize the number of revocations. To do this, we introduce restrictions in the credential to be valid. These restrictions involve the validity period and in general any parameter included in the credential Type.

We are now ready to define delegation and give an appropriate notation for this concept.

Definition 2 (Delegation Credential)
A delegation credential is a tuple (I, H, A, T) in $\mathcal{E}^* \times \mathcal{E} \times \mathcal{A} \times \mathcal{T}$ where,

- *$(\mathcal{E}, \leq)$ is a partial order set representing the possible issuers and holders of delegation credentials.*
- *$(\mathcal{A}, \leq)$ is a partial order set representing the possible attributes we consider in our systems.*
- *$(\mathcal{T}, \leq)$ is a partial order set representing the additional properties of the credentials that have to be considered when deciding if a chain of credentials is valid or not.*

The set of all possible credentials is denoted by $\mathcal{C} := \mathcal{E}^* \times \mathcal{E} \times \mathcal{A} \times \mathcal{T}$.

The meaning of $(Alice, Bob, Attribute, Type)$ is that $Alice$ issues a credential regarding the attribute $Attribute$ to Bob, and that this credential has type $Type$.

3.1 Validity Evaluations

When defining the credentials in the system, not all the possible credentials of $\mathcal{C}$ are going to be considered as valid. Instead of defining a subset of valid credentials, we define a map from $\mathcal{C}$ to the set $\{true, false\}$ in such a way that all the valid credentials will be mapped to $true$ and the non-valid to $false$. But this map should also take into account the state or context of the system: instant of time, context of entities, location of entities, etc., as this information could interfere in the credential validation process.

Let $States$ be the set of possible states of our system. Each estate $s \in States$ encodes all the relevant contextual information for making delegation and authorization decisions. In the simplest example it is reduced to the system time, so the only relevant information for a credential to be valid is the validity interval, but in a functional system, each state should include all the relevant information for determining the validity of any credential. We define then a function

$$f : \mathcal{C} \times States \rightarrow \textbf{Boolean}$$

which decides if a given instant of time (or state) is included in a certain time interval. Using this function, a credential (I, H, A, T) is valid in state s if and only if $f(\_, \_, \_, T, s) = true$. A function f like this is a *validity evaluation*.

Definition 3 (Validity Evaluation)
Let S be the set of all possible states of the system. A function

$$f : \mathcal{C} \times \mathcal{S} \rightarrow \{true, false\}$$

is a validity evaluation if and only if

$$f(I, H, A, T) = true \Longrightarrow f(I', H', A', T') = true$$

for all (I', H', A', T') where $I = I'$, $H' \leq H$, $A' \leq A$ and $T' \leq T$.

We distinguish between two sorts of evaluations, those that depend on the subjects involved in the credential, i.e. the issuer and the holder, and those that do not depend on them. We also distinguish between those functions that depend on the attribute and those that do not depend on it.

Definition 4 (Classes of validity evaluations)
*An **objective** validity evaluation is a function which depends only on the attributes and the type of the credential,*

$$f(I, A, H, T, s) = f(I', H', T, A, s) \, \forall I', H'$$

*An **universal** validity evaluation is a function which does not depend on the attributes,*
$$f(I, A, H, T, s) = f(I, H, T, A', s) \, \forall A'$$

*A **subjective** validity evaluation is a function that is not objective, i.e. depends on the subject or the issuer of the credential.*

Objective validity evaluations do not care about the issuers or holders of the delegation credentials but are concerned with the attributes issued in each credential. On the other hand, subjective validity evaluations are affected by the entities (holders and issuers) involved in the credentials.

As an example, let us think of a reputation system and suppose that the reputation of each entity is stored in the state of the system. If we take the reputation of the entities involved in a delegation credential into account in order to decide if the chain is valid or not, then we are using a subjective validity evaluation. If, on the other hand, the entities are not considered then we are using an objective validity evaluation.

Another example of validity evaluation is to use the instant of time. In this way
$$f(I, A, H, T, s) = true \text{ iff } time(s) \leq time(T)$$

where $time(\cdot)$ gives us both the information of the state regarding time and the time component of the Type. The symbol $\leq$ represents the subclass relation (partial order) of the time component of the type. This easy schema can be used with other Type components, like location. Location can be encoded using IP addresses with and ontology encoding the subnetting relation or using geographical information [9,10]

We encode the validity evaluation using RuleML [7] or SWRL [8] which is also supported by a Protege plugin. The definition of the last kind of validity evaluation in SWRL is trivial, as it only involves one subclass relation. With some complex examples we have to use RuleML.

4   Chaining Delegation Credentials

We mentioned before that unlike authorization credentials, delegation credentials can be chained to form a delegation chain. This consists of a sequence of delegation credentials concerning the same attribute and in which the issuer of a credential is the holder of the previous credential. Furthermore, in any given path, the issuer of a credential has to be lower, in the subclass relation, than the holder of the previous credential, formally:

Definition 5 (Delegation Path). *A sequence of delegation credentials* $\{C_i\}_{i=1}^n$, *where* $C_i = (I_i, H_i, A_i, T_i)$, *is a delegation path or chain for the attribute* A *if,*

1. $I_{i+1} \leq H_i$ *for all* $i \in \{1, \ldots, n\}$.
2. $A \leq A_i$ *for all* $i \in \{1, \ldots, n\}$.
3. $D \leq T_i$ *for all* $i \in \{1, \ldots, n\}$.

Where D *represents the minimal type for delegation credentials. A sequence of delegation credentials* $C := \{D_i\}_{i=1}^n$ *is a chain or path if there exists an attribute* $A \neq \emptyset$ *such as* C *is a delegation chain for* A. *The set of all delegation paths is denoted by* $\mathcal{P}$.

Condition *1*, in Definition 5, makes use of the partial order given on the set of entities. When an entity y is more specific, lower in the hierarchy, than another entity x, then y *inherits* the attributes issued to x. In the extreme situation in which the partial order is trivial, this condition is reduced to $I_{i+1} = H_i$ for all $i \in \{1, \ldots, n\}$.

Condition *2*, makes use of the partial order given on the set of attributes. When an entity x issues a credential over any attribute a, it is implicit that any other attribute a' which is more specific than a ($a' \leq a$) is also issued. Thus, we use this implicit rule to chain credentials that have some attributes in common.

Condition *3*, only establishes that all credentials in a delegation path must be delegation credentials. We use the type element D to represent the type *delegable*.

Given a path of credentials, we can map it to a single credential using a sequential operator. This operator is well defined only when the partial order sets $\mathcal{A}$ and $\mathcal{T}$ are not only partial orders but semi-lattices for the *meet* operator. In this case we take advantage of the *meet* operator for lattices to define a sequential operator for credentials.

Definition 6 (Sequential Operator). *Let* $\wedge$ *denote the meet operator of the lattices* $\mathcal{A}$ *and* $\mathcal{T}$. *Then, given two credentials* (X, Y, A, T) *and* (Y', Z, A', T') *with* $Y' \leq Y$ *we define the sequential operator as*

$$(X, Y, A, T) \wedge (Y', Z, A', T') = (X, Z, A \wedge A', T \wedge T').$$

Using this operator we give an alternative definition for credential paths

Definition 7 (Delegation Path for lattices). *Let* $\{C_i\}_{i=1}^n$ *be a sequence of delegation credentials, where* $C_i = (I_i, H_i, A_i, T_i)$ *and let* $(I_P, H_P, A_P, T_P) = C_1 \wedge C_2 \wedge \ldots \wedge C_n$. *The sequence is a delegation path or chain for the attribute* A *if,*

1. $I_{i+1} \leq H_i$ for all $i \in \{1, \ldots, n\}$.
2. $A \leq A_P$.
3. $D \leq T_P$.

Making use of the sequential operator we can map each credential path with a single credential. These credentials encode the meaning of the path and will be used when taking authorization and delegation decisions.

If we have a poset we may complete it with the special element $\emptyset$, that is defined as the minimal element in the poset, in such a way that the resulting set is indeed a semi-lattice for the meet operator. Then, we can use the previous definition with the extended posets.

Now we define the concept of valid credential path. We decide if a chain of credentials is valid or not, in a given state, using the same idea as with simple credentials. To do so, we define a *validity function* to decide whether a chain of credentials is valid or not.

Definition 8 (Validity Function). *Let $\mathcal{S}$ be the set of all possible states of the system. A function*

$$f : \mathcal{P} \times \mathcal{S} \to \{true, false\}$$

is a validity function if restricted to the domain $\mathcal{C} \times \mathcal{S}$ is a validity evaluation.

The first and simplest approach to determine if a path of credentials is valid is to check whether all the credentials of the path are valid in the state. Indeed, this is the least restrictive approach. So, we call it *LR validity function*.

Definition 9 (LR validity function). *Let $P := C_1 C_2 \ldots C_n$ be a chain of credentials. The Least Restrictive (LR) validity function is defined by,*

$$\hat{f} \equiv f_{LR} : \mathcal{P} \times \mathcal{S} \longrightarrow \{true, false\}$$
$$(P, s) \longmapsto \bigwedge_{i=1}^{n} f(C_i, s)$$

In the simple case in which, $f(C_i, s) \equiv f(\_, \_, \_, T_i, s)$, $\hat{f}$ depends only on the types of the credentials that composed the path P. As with validity evaluations, we distinguish between Objective and Subjective validity functions.

4.1 Examples of Types

We introduce here two incremental examples. We focus on the definition of the type of the credentials and on the validity evaluations and functions associated to the credentials. First of all, we define the set of States $\mathcal{S}_0$ consisting of points in time. Let define

$$\mathcal{T}_0 := \mathcal{T}_I \times \mathcal{T}_D$$

where $\mathcal{T}_I$ and $\mathcal{T}_D$ are the types defined in Section 2.3.

We define a Universal Objective validity evaluation as, $f(I, H, A, T, s) = true$ if and only if $s \in T$.

The validity function defined above is clearly universal and objective as it only depends on the type of the credentials and of course on the given state. Let us try to reduce the condition $s \in T$ to a more general condition using only the partial order. If we represent the states of S_0 as unitary intervals:

$$S_0 := \{[s, s] : s \in \mathbb{N}\}$$

then the validity function f_0 is defined as the following:

$$
\begin{aligned}
f_0 : \mathcal{P} \times S_0 &\longrightarrow \{true, false\} \\
(P, s) &\longmapsto (s \leq T)
\end{aligned}
\tag{1}
$$

Suppose we want to use a Multilevel security policy in which we define two security levels: *weak* and *strong*. Suppose that the *strong* is more restrictive than the *weak* level, so there could be credentials that are valid for the *weak* but not for the *strong* one. In this case, we should include the label *weak* in those credentials that are only valid in the *weak* level and the *strong* label in those which are valid in any level. This situation can be easily encoded using partial order. We define a new set of states, S_1, that contains the level of security of the state and a point in time.

$$S_1 := S_0 \times \{weak, strong\}$$

Analogously, we define a new type,

$$T_1 := T_0 \times \{weak, strong\}$$

that is a product of partial orders, where the partial order of $\{weak, strong\}$ is defined with the inequality $weak \leq strong$. With those definitions, we define the validity evaluation, f_1, as in Equation 1.

In those cases in which we could give a meaning to $s \leq T$ we refer to f_0 as the *canonical* validity evaluation and to $\hat{f}_0$ as the *canonical* validity function.

The last example is a subjective validity function that requires a reputation system. Suppose $r(E)$ gives us the reputation of entity E as a real number in the interval $[0, 1]$. We can define a lower bound of 0.5 for the reputation of the issuer of the first credential in the path. In this way $f(P, s) = true$ if and only if $f_{LR}(P, s) = true$, $I_1 \in \mathcal{E}^*$ and $r(I_1) \geq 0.5$.

5   Conclusions

We have defined a general mathematical framework for model delegation. Although we have used a mathematical notation, the ideas presented in this paper could have been formulated using a more common language. The use of partial orders is clearly supported by ontologies, and in particular OWL offers a subclass mechanism that is well suited to the concept of partial order. So, in practice, when we talk about partial orders, we are thinking about a simple subclass ontology. More work has to be done in order to support more complex ontologies. The other interesting concept presented in this paper is the context, which is encoded in the variable *state*. All information relevant to the system is encoded using ontologies which allows us to use rule languages such as RuleML and SWRL to reason on the delegation and authorization relationships in the system.

References

1. Isaac Agudo, Javier Lopez and Jose A. Montenegro. "A Representation Model of Trust Relationships With Delegation Extension". In 3rd International Conference on Trust Management, iTrust 2005, volume 3477 of Lecture Notes in Computer Science, pages 116 - 130. Springer, 2005.
2. Aditya Kalyanpur, Bijan Parsia, Evren Sirin, Bernardo Cuenca-Grau and James Hendler, "Swoop: A Web Ontology Editing Browser", Journal of Web Semantics Vol 4(2), 2005
3. D.F. Ferraiolo, D.R. Kuhn and R. Chandramouli, "Role Based Access Control", Artech House, 2003.
4. C. Ellison, B. Frantz, B. Lampson, R. Rivest, B. Thomas and T. Ylonen "SPKI Certificate Theory", *RFC 2693*, 1999.
5. Holger Knublauch, Ray W. Fergerson, Natalya F. Noy and Mark A. Musen "The Protege OWL Plugin: An Open Development Environment for Semantic Web Applications" Third International Semantic Web Conference - ISWC 2004, Hiroshima, Japan, 2004.
6. S. Bechhofer et al., "OWL Web Ontology Language Reference". 2004.
7. Boley, H., "The Rule Markup Language: RDF-XML Data Model, XML Schema Hierachy, and XSL Transformations", Invited Talk, INAP2001, Tokyo, Springer-Verlag, LNCS 2543, 5-22, 2003.
8. "SWRL: A Semantic Web Rule Language Combining OWL and RuleML". W3C Member Submission. 21-May-2004.
9. B. Purevjii, T. Amagasa, S. Imai, and Y. Kanamori. "An Access Control Model for Geographic Data in an XML-based Framework". In Proc. of the 2nd International Workshop on Information Systems Security (WOSIS), 2004, pages 251–260.
10. V. Atluri and P. Mazzoleni. "A Uniform Indexing Scheme for Geo-spatial Data and Authorizations". In Proc. of the Sixteenth Conf. on Data and Application Security, IFIP TC11/WG11.3, Cambridge, UK, 2002, pages 207–218.

Adding Support to XACML for Dynamic Delegation of Authority in Multiple Domains

David W Chadwick, Sassa Otenko, and Tuan Anh Nguyen

University of Kent, Computing Laboratory, Canterbury, Kent, CT2 7NF
d.w.chadwick@kent.ac.uk, o.otenko@kent.ac.uk, tn32@kent.ac.uk

Abstract. In this paper we describe how we have added support for dynamic delegation of authority that is enacted via the issuing of credentials from one user to another, to the XACML model for authorisation decision making. Initially we present the problems and requirements that such a model demands, considering that multiple domains will typically be involved. We then describe our architected solution based on the XACML conceptual and data flow models. We also present at a conceptual level the policy elements that are necessary to support this model of dynamic delegation of authority. Given that these policy elements are significantly different to those of the existing XACML policy, we propose a new conceptual entity called the Credential Validation Service (CVS), to work alongside the XACML PDP in the authorisation decision making. Finally we present an overview of our first specification of such a policy and its implementation in the corresponding CVS.

Keywords: XACML, Delegation of Authority, Credentials, Attributes, Policies, PDP.

1 Introduction

XACML is an OASIS standard for authorisation decision making. Many people are starting to experiment with it in their applications e.g. [11, 12]. Some of its benefits include: a flexible attribute based authorisation model, where access control decisions can be made based on the attributes of the subject, the action and the target; a comprehensive way of specifying conditions, so that arbitrarily complex conditions can be specified; and the support for obligations.

However, one of the current drawbacks of using XACML is that it does not support dynamic delegation of authority. A delegate is defined as "A person authorized to act as representative for another; a deputy or an agent" [1]. Delegation of authority is the act of one user with a privilege giving it to another user (a delegate), in accordance with some delegation policy. A delegation tree may thus be created, starting from the root user who has the privilege initially, to the users at the leaves of the tree who end up with the authority to assert the delegated privilege, but cannot delegate it themselves. Non leaf nodes in the tree are authorities (or administrators) with permission to delegate, but may or may not be able to assert the privilege themselves (according to the delegation policy). We differentiate between static and dynamic delegation of authority, in that static delegation of authority is when the non leaf nodes of the delegation tree are configured into software (or policy)

H. Leitold and E. Markatos (Eds.): CMS 2006, LNCS 4237, pp. 67–86, 2006.

prior to user access i.e. the shape of the delegation tree is known from the start, and no new non-leaf nodes can be created without reconfiguring the software (or policy). Dynamic delegation of authority is when only the root user and delegation policy are configured into the software prior to user access, and users may dynamically delegate authority to other users as and when they wish. In this case the delegation tree is created dynamically as one user delegates to another, and new leaf (and non-leaf) nodes are created spontaneously.

A responsive authorisation infrastructure that can cater for rapidly changing dynamic environments should be able to validate the privileges given to any of the users in a dynamically created delegation tree, even though the actual tree is not known when the authorisation policy is written and fed into the policy decision point (PDP). This requires the authorisation policy to be supplemented with a delegation policy that will state how the delegation tree is to be constrained. As long as a user's credential falls within the scope of the delegation tree then it is considered valid, if it falls outside the tree, and thus outside the delegation policy, it is not. The purpose of the current research was to add dynamic delegation of authority to an authorisation infrastructure that contains an XACMLv2 PDP (or in fact any PDP that bases its access control decisions on the attributes of subjects), without changing the XACMLv2 PDP or its policy[1].

We assume that privileges can be formulated as attributes and given to users. An important point to clarify at the outset is the difference between an attribute and a credential (i.e. authorisation credential). An attribute is a property of an object[2]; a credential is a *statement* or *assertion* about an attribute (in particular, a credential must state: what the attribute is, who the attribute belongs to, who says so (i.e. who is the credential issuer), and any constraints on its validity). Because attributes of an entity do not always exist as part of the entity, they are often stored or transferred as separate stand alone credentials. In this paper we are concerned with dynamic delegation of authority from one user to another by the use of credentials. One important feature of a credential is that it requires validation before the user can be attributed with the asserted property.

The rest of this paper is structured as follows. Section 2 describes the problems that need to be addressed when creating an infrastructure to support dynamic delegation of authority between multiple domains, and this leads to various requirements being placed on any proposed solution. Section 3 describes the new conceptual credential validation service (CVS) that is proposed to resolve the problems and requirements described in Section 2. Section 4 briefly describes the XACMLv2 infrastructure. Section 5 discusses how the CVS could be incorporated into the XACML infrastructure. Section 6 describes our implementation of a CVS. Section 7 concludes, and looks at possible future work in this area.

---

[1] Note that this research started whilst XACMLv2 was still under construction, when it was known that XACMLv2 would not support dynamic delegation of authority. This was one of the reasons for not proposing changes to XACMLv2. Work is currently underway to add administration and delegation to XACML v3 [18], but this is complementary to the work described here.

[2] Dictionary.com defines an attribute as "A quality or characteristic inherent in or ascribed to someone or something".

2 Problem and Requirement Statements

The underlying model used for dynamic delegation of authority in multiple domains is an enhancement of the basic XACMLv2 model (see later). In this enhanced model a user (subject) is dynamically given a set of attributes by one or more dynamically created attribute authorities (AAs) in one or more domains, and these attributes are presented (pushed) to or obtained (pulled) by the PDP as a set of credentials (usually in the form of attribute assertions digitally signed by the AAs). The PDP makes its access control decisions based on its policy, the validated set of subject attributes, the target and environmental attributes and the parameters of the user's request. Below are a set of issues that need to be addressed in such a model.

1. **Valid vs. Authentic Credentials.** The first thing to recognise is the difference between an *authentic* credential and a *valid* credential. An *authentic* credential, from the perspective of authorisation decision making, is one that has been received exactly as it was originally issued by the AA. It has not been tampered with or modified. Its digital signature, if present, is intact and validates as trustworthy meaning that the AA's signing key has not been compromised, i.e. his public key (certificate) is still valid. A *valid* credential on the other hand is one that is trusted by the PDP's policy for authorisation decision making. In order to clarify the difference, an example is the paper money issued by the makers of the game Monopoly. This money is authentic, since it has been issued by the makers of Monopoly. The money is also valid for buying houses on Mayfair in the game of Monopoly. However, the money is not valid if taken to the local supermarket.

2. **Credential validity is determined by target domain.** The above discussion leads onto the second problem that needs to be addressed in any solution, and this is that there are potentially *multiple domains* within an authorisation infrastructure. There are issuing domains, which issue credentials, and target domains that consume credentials. The PDP is part of the target domain, and as such it must use the policy of the target domain to decide whether a credential is to be trusted or not i.e. is valid or not. So the validity of an authorisation credential is ultimately determined by the (writer of the) PDP policy. A valid credential is a credential that is trusted by the consumer of the credential.

3. **Multiple trusted credential issuers.** In any system of any significant size, there will be multiple credential issuers. Some of these will be trusted by the target domain, others will not be. Thus the system must be capable of differentiating between trusted and untrusted issuers, and of dynamically obtaining this information from somewhere. (In point 4 below we propose to use roots of trust.) Different target domains in the same system may trust different issuers, and therefore the PDPs must be capable of being flexibly configured via their policies to say which issuers are trusted and which are not. For example, in the physical world of shopping with credit cards, there are several issuers such as Amex and Visa. Some shopkeepers accept (trust) both issuers, others only trust one of them. It is their (the target domain's) decision which card issuers to trust.

4. **Identifying roots of trust.** Point 3 above leads us to conclude that the PDP must be configured, in an out of band trusted way, with at least one (authorization) root

of trust and it is from this root (or roots) of trust that all credentials must be validated in order to be trusted. A root of trust must be a single entity identified directly or indirectly by its public key[3], since this key will be used to validate the signed credentials that are received. Note that it is not possible to refer to a *root of trust* through its set of assigned attributes, e.g. anyone with a project manager attribute and company X attribute, since these attributes may identify several candidate roots, and may be issued by several attribute authorities, in which case it wont be known who to trust. This implies that a higher authority is the real root of trust, the one who issues the set of attributes that can be trusted.

5. **The role of the Issuer's policy.** Most issuers will have an Issuing Policy, to say who is allowed to issue which credentials to which users, and what constraints are placed on their use. This policy will include the delegation policy of the issuer. Consequently there will be constraints on which credentials are deemed to be valid within and without the issuing domain. However, the target domain may choose to ignore these constraints and trust (treat as valid) credentials which the issuer deems to be invalid. A well known example in the physical world concerns supermarkets who issue their own discount coupons. These coupons state quite clearly that they are only valid for use in supermarkets owned by the issuer. However, it is often the case that a different brand of supermarket will accept these discount coupons as a way of enticing the other supermarkets' customers to come and shop in their own supermarket. Thus the PDP must have a way of either conforming to or overriding the issuer's policy. If a target domain chooses to ignore the issuer's policy, then it is liable for any losses incurred by this. The issuer cannot be held responsible for targets that ignore its Issuing Policy.

6. **Obtaining the Issuing Policy.** In a multi-domain system, the target domain may not be aware of the issuing domain's Issuing Policy, unless it is explicitly placed into the issued credentials. If the complete Issuing Policy is not explicitly placed in the issued credentials, but the target domain still wishes to enforce it and only treat as valid those credentials that the issuer says are valid, then the target's PDP will need to infer or be configured with the issuer's Issuing Policy. For example, in SPKI [7], a credential is marked as being infinity delegatable or not, and does not contain any other details of the Issuing Policy, such as who is entitled to be delegated the privilege. Thus unless a delegatable credential explicitly contains restrictions, or out of band means are used to transfer them, the target PDP will infer than anyone is entitled to be delegated this credential.

7. **Pulling credentials.** The PDP may not have all the credentials it needs in order to validate the credential(s) presented by the user, e.g. if only the leaf credential in a delegation tree is presented, but none of the non-leaf credentials are presented. In the most extreme case the user may not present any credentials at all. For example, when a user logs into a portal and the portal displays only the services this user is allowed to see, the portal has, unknown to the user, retrieved the user's credential(s) from a repository in order to determine which services to

---

[3] When an X.509 conformant PKI is used which already has its own configured CA root public keys, the globally unique name of the subject in the PKI certificate can be used to refer to the authorization root of trust, instead of the public key in the certificate, in which case the subject will be trusted regardless of which public/private key pair it is currently using.

display. There is thus a strong requirement for the PDP (or a component of it) to be able to pull the user's credentials before making the access control decision.

8. **Discovering credential locations.** The user's credentials may be stored and/or issued in a variety of places, for example, each AA may store the attributes or the credentials it issues in its own repository. One could always mandate that the user collects together the credentials he wants to use, before attempting to gain access to a resource e.g. as in the VOMS model [13]. This model has its merits in some cases, but it is not always very user friendly. In fact, in some cases, the user may not be aware what credentials have actually been issued to him – he might only know what services he is allowed to access, as in the portal example given above. In the general case there is no absolute requirement for the user to know what credentials have been issued to him or where they are stored. Thus the PDP must be capable of contacting different repositories/AAs in order to pull the user's credentials prior to making its access control decision.

9. **Multiple user identities.** If the user is known by different identities to the different AAs, then there must be a way for the user to use these mixed credentials in the same session. The GridShib project currently uses a mapping table to convert between X.509 PKI identities and Shibboleth identity provider identities [14]. But a more flexible approach is needed, in which the user may determine which set of credentials are to be used in a given session and the PDP can prove the user's right to assert each one. We propose one solution to this in [20].

10. **Multiple credential formats.** Following on from above, the user's credentials may be stored in different formats in the different repositories, and presented to the PDP in different ways, e.g. as signed SAML assertions [2], as X.509 attribute certificates [3], as Shibboleth encoded attributes [4] etc. The PDP (or a component of it) therefore needs to be able to decode and handle credentials in different formats.

11. **Hierarchies of attributes.** The attributes may form some sort of hierarchy, for example in accordance with the NIST $RBAC_1$ specification [5], in which the superior attributes (or roles) inherit the privileges of the subordinate roles. The PDP needs to be aware of this hierarchy when validating the credentials. For example, if a superior role holder delegates a subordinate role to another user, then the PDP needs to know if this delegation is valid or not, given that the attributes are different. Furthermore some of the attributes known to the PDP won't form a hierarchy. Therefore the PDP needs to be able to cater for multiple disjoint attribute hierarchies.

12. **Constraining credential validity.** Only part of an authentic credential might be valid in a target domain. For example, a credential might contain multiple attributes but the target domain only trusts the issuer to issuer a subset of the enclosed attributes.

13. **Known and unknown attributes.** As federations between organisations become more common, and dynamic VOs become more feasible, managers will realise the need to define a common set of attributes that can be understood between domains. The US academic community realised this some time ago, and this led to the definition of EDU person [6], which is a collection of standard attribute types. However, once organisations start to issue standard attributes, a PDP will

need to be able to differentiate between which standard attributes are valid (trusted) and which are not. For example, suppose most organisations in the world issue a standard Project Manager attribute to their project managers. In a VO between organisations A and B, the PDP in organisation B might only want to trust the Project Manager attributes issued by itself, and not those issued by organisation A (or by C or D or any other organisation). Or alternatively it might wish to downgrade those issued by organisation A and treat them as being equivalent to a guest user attribute. Or it might decide to trust the project managers from A as being equal to its own project managers. The PDP's policy needs to flexible enough to cater for all these requirements, including the ability to perform attribute mappings.

3 Architecting a Solution

Given the problem statements and various requirements from above, one can see that some new functional requirements have been placed on the PDP. Consequently, we propose a new conceptual component called a Credential Validation Service (CVS), whose purpose is to perform the new functionality. In essence the purpose of the CVS is to validate a set of credentials for a subject, issued by multiple dynamic attribute authorities from different domains, according to the local policy, and return a set of valid attributes. How this conceptual component is merged into the XACML infrastructure will be described later. The rationale for making the CVS a separate component from the XACML PDP are several. Firstly, its purpose is to perform a distinct function from the PDP. The purpose of the PDP is to answer the question "given this access control policy, and this subject (with this set of valid attributes), does it have the right to perform this action (with this set of attributes) on this target (with this set of attributes)" to which the answer is essentially a Boolean, Yes or No[4]. The purpose of the CVS on the other hand is to perform the following "given this credential validation policy, and this set of (possibly delegated) credentials, please return the set of valid attributes for this entity" to which the answer will be a subset of the attributes in the presented credentials. Secondly, the XACML language is incapable of specifying credential chains. This is because subjects and attribute issuers are identified differently in the language, hence it is not possible to chain delegated credentials together.

When architecting a solution there are several things we need to do. Firstly we need a trust model that will tell the CVS which credential issuers and policy issuers to trust. Secondly we need to define a credential validation policy that will control the trust evaluation of the credentials. Finally we need to define the functional components that comprise the CVS.

3.1 The Trust Model

The CVS needs to be provided with a trusted credential validation policy. We assume that the credential validation policy will be provided by the Policy Administration

---

[4] XACML also supports other answers: indeterminate (meaning an error) and not applicable (meaning no applicable policy), but these are other forms of No.

Point (PAP), which is the conceptual entity from the XACML specification that is responsible for creating policies. If there is a trusted communications channel between the PAP and the CVS, then the policy can be provided to the CVS through this channel. If the channel is not trusted, or the policy is stored in an intermediate repository, then the policy should be digitally signed by a trusted policy author, and the CVS configured with the public key (or distinguished name if X.509 certificates are being used) of the policy author. In addition, if the PAP or repository, has several different policies available to it, to be used at different times, then the CVS needs to be told which policy to use. In this way the CVS can be assured of being configured with the correct credential validation policy. All other information about which sub policies, credential issuers and their respective policies to trust can be written into this master credential validation policy by the policy author.

In a distributed environment we will have many issuing authorities, each with their own issuing policies provided by their own PAPs. If the policy author decides that his CVS will abide by these issuing policies there needs to be a way of securely obtaining them. Possible ways are that the CVS could be given read access to the remote PAPs, or the remote issuing authorities could be given write access to the local PAP, or the policies could be bound with their issued credentials and obtained dynamically during credential evaluation. Whichever way is used, the issuing policies should be digitally signed by their respective issuers so that the CVS can evaluate their authenticity.

The policy author may decide to completely ignore all the issuer's policies (see section 2 point 5), or to use them in combination with his own credential validation (CV) policy, or to use them in place of his own policy. Thus this information (or policy combining rule) needs to be conveyed as part of the CV policy.

3.2 The Credential Validation Policy

The CVS's policy needs to comprise the following components:

- a list of trusted credential issuers. These are the issuers in the local and remote domains who are trusted to issue credentials that are valid in the local domain. They are the roots of trust. This list is needed so that the signatures on their credentials and policies can be validated. Therefore the list could contain the raw public keys of the issuers or it could refer to them by their X.500 distinguished names or their X.509 public key certificates.

- the hierarchical relationships of the various sets of attributes. Some attributes, such as roles, form a natural hierarchy. Other attributes, such as file permissions might also form one e.g. *all* permissions is superior to *read, write* and *delete*; and *write* is superior to *append* and *delete*. When an attribute holder delegates a subordinate attribute to another entity, the credential validation service needs to understand the hierarchical relationship and whether the delegation is valid or not.

- a description (schema) of the valid delegation trees. This delegation policy component describes how the CVS can determine if a chain of delegated credentials and/or policies falls within a trusted tree or not. This is a rather complex policy component, and there are various ways of describing delegation trees [3, 9] with no widely accepted standard way. The essential elements should specify who is allowed to be in the tree (both as an issuer and/or a subject), what constraints apply, and what properties (attributes) they can validly have (assert) and delegate.

- a linking of trusted issuers to delegation trees. This is not necessarily a one to one mapping. Several trusted issuers may be at the root of the same delegation tree, or one issuer may be at the root of several delegation trees.

- the acceptable validity constraints of the various credentials (e.g. time constraints or target constraints). Consider for example time constraints. An issuer gives each issued credential a validity period, which may range from fairly short (e.g. minutes) to very long (e.g. several years). The primary reason for issuing short lived certificates (for other than intrinsically short lived permissions) is so that they do not need to be revoked, and therefore the relying party does not need to consult revocation lists, white lists, or OCSP servers etc. In the case of relatively long lived credentials, the policy author may have his own opinion about which credentials to trust, from a chronological perspective, and therefore may wish to place his own additional time constraints on remotely issued credentials. For example, a plumber may have a "certified plumber" credential, which is valid for 10 years from the date of issue. He may be required to pass a competence test every ten years to prove that he is conversant with the latest technology developments and quality standards before the credential is renewed. However, in the target domain, the policy author may decide that he does not want to accept anyone with a credential that is newer than one year old, due to insufficient experience on the job, or is more than 8 years old, due to doubts about competencies with the latest technologies. Consequently the CVS must be told how to intersect the validity constraints on the credential with those in the author's policy.

- finally, we need a disjunctive/conjunctive directive, to say whether for each trusted issuer and delegation tree, only the issuer's issuing and delegation policy should take effect, or only the author's policy should take effect, or whether both should take effect and valid credentials must conform to both policies.

Note that when delegation of authority is not being supported, the above policy can still be used in simplified form where a delegation tree reduces to a root node that describes a set of subjects. In this case the CV policy now controls which trusted issuers are allowed to assign which attributes to which groups of subjects, along with the various constraints and disjunctive/conjunctive directive.

XACMLv2 [8] is not a suitable instrument to express Credential Validation Policies but neither is the current working draft of XACMLv3 [18]. An important requirement for multi-domain dynamic delegation is the ability to accept only part of an asserted credential. This means that the policy should be expressive enough to specify what is the maximum acceptable set of attributes that can be issued by one Issuer to a Subject, and the evaluation mechanism must be able to compute the intersection of this with those that the Subject's credential asserts. The approaches used by XACML can only state that the asserted set of attributes or policies is fully accepted, or fully rejected. In [18] the delegation is deemed to be valid if the issuer of the delegated policy could have performed the request that the policy grants to the delegatee. We think this is a serious deficiency, which lies at the core of the XACML policy evaluation process.

We think it is a limitation on an independent issuing domain to have to take into account all the policies that the validating domain supports, so that only fully acceptable sets of credentials or policies can be issued to its subjects. Our model is based on full independence of the issuing domain from the validating domain. So in general it is impossible for a validating domain to fully accept an arbitrary set of credentials since the issuing and validating policies will not match. It is not possible for the issuing domain to tell in advance in what context a subject's credentials will be used (unless new credentials are issued every time a subject requests access to a resource) so it is not possible to tell in advance what validation policy will be applied to them.

Having identified this problem, we propose a solution that uses a non-XACML based credential validation policy first, and an XACML policy next, with the validated delegated attributes.

3.2.1 Formal Credential Validation Policy

We define a Credential Validation Policy as an unordered set of tuples $<S, I, C, E>$, where S is a set of Subjects to whom any Issuer from set I can assign at most a set of Credentials C, but only if any of the conditions in set E holds true:

$$CVP = \{<S, I, C, E>\}$$

We define the Credential Validation process as a process of obtaining a subset of valid credentials V, given an asserted set of credentials c, issued by issuer i to the subject s, if condition e holds true at the time of evaluation:

$$V = \{ c \cap C \mid c \cap C \neq \varnothing, s \subseteq S, i \subseteq I, e \subseteq E, <S, I, C, E> \subseteq CVP \}$$

Note that in XACML the only possible evaluation of a Credential Validation process is:

$$V = \{ c \mid c \subseteq C, s \subseteq S, i \subseteq I, e \subseteq E, <S, I, C, E> \subseteq CVP \}$$

Further, we define a dynamic delegation process as a process of obtaining a set R of Credential Validation rules for intermediate issuers, i.e. the issuers on the path from the policy writer to the end user, where the intermediate issuer s is issued a set of Credentials c by a higher level issuer i, subject to condition e and a constraint on subject domain d:

$$R_s = \{ <d \cap S \backslash s, s, c \cap C, e> \mid c \cap C \neq \varnothing, s \subseteq S, i \subseteq I, e \subseteq E,$$

$$<S, I, C, E> \subseteq CVP \cup R_i \}$$

Thus the issuer i can allow the issuer s to delegate a subset of his own privileges to a subset of his own set of subjects, subject to the condition e being stricter than that imposed on i.

Note the recursive nature of the process - the tuple $<S, I, C, E>$ must belong to the CVP or to the set of valid rules for issuer i. Note also that loops in the delegation are prohibited by excluding the holder of the rule from the set of possible subjects.

XACML currently lacks the expressiveness for deriving new Credential Validation rules given the set of existing rules and valid credentials.

3.3 The CVS Functional Components

Figure 1 illustrates the architecture of the CVS function and the general flow of information and sequence of events. First of all the service is initialised by giving it

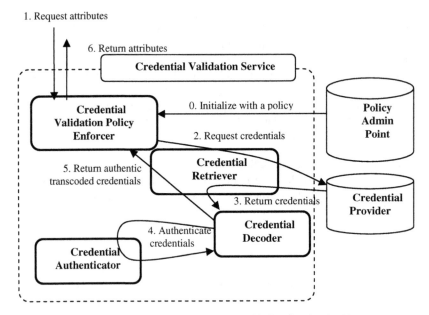

Fig. 1. Data Flow Diagram for Credential Validation Service Architecture

the credential validation policy (step 0). Now the CVS can be queried for the valid attributes of an entity (step 1). Between the request for attributes and returning them (steps 1 and 6) the following events may occur a number of times, as necessary i.e. the CVS is capable of recursively calling itself as it determines the path in a delegation tree from a given node to a root of trust. The Policy Enforcer requests credentials from a Credential Provider (step 2). When operating in credential pull mode, the credentials are dynamically pulled from one or more remote credential providers (these could be AA servers, LDAP repositories etc.). The actual attribute request protocol (e.g. SAML or LDAP) is handled by a Credential Retriever module. When operating in credential push mode, the CVS client stores the already obtained credentials in a local credential provider repository and pushes the repository to the CVS, so that the CVS can operate in logically the same way for both push and pull modes. After credential retrieval, the Credential Retriever module passes the credentials to a decoding module (step 3). From here they undergo the first stage of validation – credential authentication (step 4). Because only the Credential Decoder is aware of the actual format of the credentials, it has to be responsible for authenticating the credentials using an appropriate Credential Authenticator module. Consequently, both the Credential Decoder and Credential Authenticator modules are encoding specific modules. For example, if the credentials are digitally signed X.509 attribute certificates, the Credential Authenticator uses the configured X.509 PKI to validate the signatures. If the credentials are XML signed SAML attribute assertions, then the Credential Authenticator uses the public key in the SAML assertion to validate the signature. The Credential Decoder subsequently discards all unauthentic credentials – these are ones whose digital signatures are invalid. Authentic credentials are decoded and transformed into an implementation specific local format that the Policy Enforcer is able to handle (step 5).

The task of the Policy Enforcer is to decide if each authentic credential is valid (i.e. trusted) or not. It does this by referring to its Credential Validation policy to see if the credential has been issued by a root of trust or not. If it has, it is valid. If it has not, the Policy Enforcer has to work its way up the delegation tree from the current credential to its issuer, and from there to its issuer, recursively, until a root of trust is located, or no further issuers can be found (in which case the credential is not trusted and is discarded). Consequently steps 2-5 are recursively repeated until closure is reached. Remember that in the general case there are multiple credential providers, who each may have their own Issuing Policies, which may be adhered to or ignored by the Policy Enforcer according to the CV policy. There are also issues of height first or breadth first upwards tree walking, or top-down vs. bottom-up tree walking. These are primarily implementation rather than conceptual issues, as they effect performance and quality of service, and so we will address them further in Section 6 where we describe our implementation of a CVS.

The proposed architecture makes sure that the CVS can:

- Retrieve credentials from a variety of physical resources
- Decode the credentials from a variety of encoding formats
- Authenticate and perform integrity checks specific to the credential encoding format

All this is necessary because realistically there is no way that all of these will fully match between truly independent issuing domains.

4   The XACML Model

Figure 2 shows the overall conceptual set of interactions, as described in XACMLv2 [8]. The PDP is initially loaded with the XACML policy prior to any user's requests being received (step 1). The user's access request is intercepted by the PEP (step 2), is authenticated, and any pushed credentials are validated and the attributes extracted (note that this is not within the scope of the XACML standard). The request and user attributes (in local format) are forwarded to the context handler (step 3), which may ask the PIP for additional attributes (steps 6 to 8) before passing the request to the PDP (step 4). If the PDP determines from the policy that additional attributes are still needed, it may ask the context handler for them (step 5). Optionally the context handler may also forward resource content (step 9) along with the additional attributes (step 10) to the PDP. The PDP makes a decision and returns it via the context handler (step 11) to the PEP (step 12). If the decision contains optional obligations they will be enforced by the obligations service (step 12).

As can be seen from Figure 2, XACMLv2 currently has nothing to say about credentials or how they are validated.

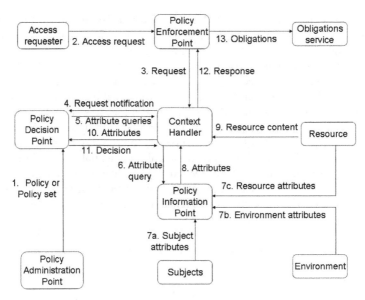

Fig. 2. Data Flow Diagram for XACML Architecture

5   Incorporating the CVS into XACML

Figure 3 shows the ways in which the CVS could be incorporated into the XACML model. The CVS could be an additional component called by either the PEP (step 101) or the context handler (step 103), or it could completely replace the PIP (step 6)

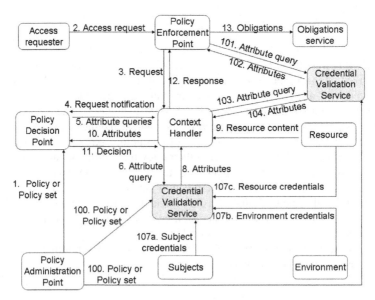

Fig. 3. Incorporating the CVS into XACML

in which case the Subject would now send credentials to the PIP/CVS rather than attributes (step 107a).

The advantages of having the CVS called by the PEP, is that existing XACMLv2 implementations do not need to change. The PEP optionally passes a bag of credentials to the CVS (push mode), the CVS fetches all or more credentials as needed (pull mode), and returns a set of valid subject attributes to the PEP, which it can pass to the existing XACML context handler. The disadvantage of this model is that each application will need to be modified in order to utilise the CVS, since the PEP is an application dependent component. Note that this model, when operating in push mode only, with no credential retrievals, is similar to that being proposed by the WS-Trust specification, in which the Security Token Service (STS) operates as a token validation service [10]. However, the STS has no equivalent functionality of the CVS operating in credential pull mode.

The advantage of having the CVS called by the context handler is that existing applications i.e. the PEP, may not need to change. The only change that is needed is to the context handler component of XACML implementations. Depending upon its interface, PEPs may not need to change at all. Support for multiple autonomous domains that each support delegation of authority can be added to applications without the application logic needing to change. Only a new credential validation policy is needed. Credentials that were previously invalid (because they had been delegated) would now become valid, once the appropriate policy is added to the PAP.

The advantage of replacing the PIP by the CVS, is that we have the opportunity of using digitally signed credentials for constructing target attributes and environmental attributes as well as subject attributes. For example, time may be obtained from a secure time stamping authority as a digitally signed credential (step 107b), and validated according to the CV policy. This is our favoured approach.

The disadvantage of the last two approaches is that incorporating the CVS inside the policy evaluator introduces transforms to the request context that are invisible to the PEP. At the current time we do not know which approach will eventually be favored.

Note that the integration scenarios do not affect the implementation of the CVS, which is explained in the next section.

6 Implementing the CVS

There are a number of challenges involved in building a fully functional CVS that is flexible enough to support the multiple requirements outlined in section 2. Firstly we need to fully specify the Credential Validation Policy, including the rules for constructing delegation trees. Then we have to engineer the policy enforcer with an appropriate algorithm that can efficiently navigate the delegation tree and determine whether a subject's credentials are valid or not.

6.1 Credential Validation Policy

We have implemented our CV policy in XML, according to our own specified DTD/schema, shown in Appendix 1. Most components of the policy are relatively straightforward to define, apart from the delegation tree. We have specified the list of trusted credential issuers by using their distinguished names (DNs) from their X.509 public key certificates. We chose to use distinguished names rather than public keys for two reasons. Firstly, they are easier for policy writers to understand and handle, and secondly it makes the policy independent of the current key pair in use by a trusted issuer.

The attribute hierarchies are specified by listing superior-subordinate attribute value pairs. There can be multiple independent roots, and attribute values can be independent of any hierarchy if so wished.

In our first implementation, we have specified a delegation tree as a name space (domain) and a delegation depth. Anyone in the domain who is given a credential may delegate it to anyone else in the same domain, who in turn may delegate it to anyone else in the same domain until the delegation depth is reached. We currently use X.500/LDAP distinguished names to define the domains. This format allows the policy writer to define a domain using included and excluded subtrees and so construct any arbitrary LDAP subtree in which the delegates must belong. This name form is already used in the PKI world, for example in the name constraints X.509 extension [3]. Furthermore since we refer to the credential subjects and issuers by their LDAP DNs, it was natural to constrain who could be in a delegation tree by referring to them by their DNs. Another constraint that we place on a delegation tree is that the same attribute (or its subordinate) must be propagated down the tree, and new unrelated attributes cannot be introduced in the middle of a delegation tree. We recognise that a more flexible approach will be to define delegation trees by referring to the attributes of the members rather than their distinguished names, as for example is used by Bandmann et al [9]. Their delegation tree model allows a policy writer to specify delegation trees such as "anyone with a head of department attribute may delegate a project manager attribute to any member of staff in the department". This is a future planned enhancement to our work.

Finally, the policy links the trusted issuers to the delegation domains and the attributes that each issuer is trusted to issue, along with any additional time/validity constraints that are placed on the issued credentials. (The constraints have not been shown in the schema.)

In our current implementation we do not pass the full Issuing Policy along with the issued credential, we only pass the tree *depth* integer. Therefore the CVS does not know what the issuer's intended name space is (we assume that the credential issuing software will enforce the Issuing Policy on behalf of the relying party). The CV policy writer is free to specify his own (more restrictive) name space for the delegation tree, or to specify no name space restrictions and allow a credential holder to effectively delegate to anyone. The only way to (partially) enforce the Issuing Policy in our current implementation is to repeat the issuer's name space in the delegation tree of the CV policy, and to assume that no further restrictions are placed by the issuer on any particular delegate. A future planned enhancement is to carry the Issuing Policy in each issued credential, and to allow the CV policy writer to enforce it, or overwrite it with his own policy, or force conformance to both. In this way a more sophisticated delegation tree can be adhered to.

6.2 Delegation Tree Navigation

Given a subject's credential, the CVS needs to create a path between it and a root of trust, or if no path can be found, conclude that a credential cannot be trusted. There are two alternative conceptual ways of creating this path, either top down, also known as backwards [3, 17] (i.e. start at a root of trust and work down the delegation tree to all the leaves until the subject's credentials are found) or bottom up, also known as forwards (i.e. start with the subject's credential and work up the delegation tree until you arrive at its root). Neither approach is without its difficulties. Either way can fail if the credentials are not held consistently – either with the subject or the issuer. As Li et al point out [17], building an authorisation credential chain is more difficult than building an X.509 public key certificate chain, because in the latter one merely has to follow the subject/issuer chain in a tree, whereas in the former, a directed graph rather than a tree will be encountered. Graphs may arise for example when a superior delegates some privileges in a single credential that have been derived from two of more credentials that he possesses, or when attribute mappings occur between different authorities. Even for the simpler PKI certificate chains, there is no best direction for validating them. SPKI uses the forwards chaining approach [15]. As Elley et al describe in [16], in the X.509 model it all depends upon the PKI trust model and the number of policy related certificate extensions that are present to aid in filtering out untrusted certificates. Given that our delegation tree is more similar to a PKI tree, and that we do not have the policy controls to filter the top/down (backwards) approach, and furthermore, we support multiple roots of trust so in general would not know where to start, then the top down method is not appropriate.

There are two ways of performing bottom up (forwards) validation, either height first in which the immediately superior credential only is obtained, recursively until the root is reached, or breadth first in which all the credentials of the immediate superior are obtained, and then all the credentials of their issuers are obtained recursively until the root or roots are reached. The latter approach may seem counter-intuitive, and certainly is not sensible to perform in real time in a large scale system, however a variant of it may be necessary in certain cases, i.e. when two or more superior credentials have been used to create a subordinate one, or when a superior possess multiple identical credentials issued by different authorities. Furthermore, given that in our federation model described in section 2 we allow a user to simply authenticate to a gateway and for the system to determine what the user is authorised to do (the credential pull model), the first step of the credential validation process is to fetch all the credentials of the user. This is performed by the Credential Retriever in Figure 1. Thus if the CVS recursively calls itself, the breadth first approach would be the default tree walking method. Thus we have to add a tree walking directive to the credential validation method, which can be set to breadth first for the initial call to the CVS, and then to height first for subsequent recursive calls that the CVS makes to itself.

In order to efficiently solve the problem of finding credentials, we add a pointer in each issued credential that points to the location of the issuer's credential(s) which are superior to this one in the delegation tree. This pointer is similar to the AuthorityInformationAccess extension defined in [19]. Although this pointer is not essential in limited systems that have a way of locating all the credential repositories, in the general case it is needed.

In the case of relatively long lived credentials, we envisage that a background task could be run when the system is idle, that works its way down the delegation trees from the roots, in a breadth first search for credentials, validates them against the CV policy, and caches the valid attributes for each user for a configuration period of time that is approximately equal to the revocation period. Then when a user attempts to access a resource, the CVS will be able to give much faster responses because the high level branches of the delegation tree will have already been validated.

7 Conclusions and Future Work

Providing XACML with support for dynamic delegation of authority that is enacted via the issuing of credentials from one user to another, is a non-trivial task to model and engineer. In this paper we have presented the problems and requirements that such a model demands, and have architected a solution based on the XACML conceptual and data flow models. We have also presented at a conceptual level the policy elements that are necessary to support this model of dynamic delegation of authority. Given that these policy elements are significantly different to those of the existing XACMLv2 policy, and that the functionality required to evaluate this policy is significantly different to that of the existing XACML PDP, we have proposed a new conceptual entity called the Credential Validation Service, to work alongside the PDP in the authorisation decision making. The advantages of this approach are several. Firstly the XACML policy and PDP do not need to change, and support for dynamic delegation of authority can be phased in gradually. The exact syntax and semantics of the new policy elements can be standardised with time, based on implementation experience and user requirements. We have presented our first attempt at defining and implementing such a policy, and now have an efficient implementation that supports dynamic delegation of authority. A live demonstration is available at http://sec.cs.kent.ac.uk/dis.html.

Future work will look at supporting more sophisticated delegation trees and schema, and enforcing (or ignoring) Issuing Policies in target domains by passing the full policy embedded in the issued credentials. We also plan to incorporate additional policy elements in the delegation trees, such as attribute mappings of the kind described in [18].

Acknowledgments. We would like to thank the UK JISC for supporting this work under the research project entitled "Dynamic Virtual Organisations in e-Science Education (DyVOSE)".

References

1. See http://dictionary.reference.com/search?q=delegate
2. OASIS. "Assertions and Protocol for the OASIS Security Assertion Markup Language (SAML) V2.0", 15 January 2005

3. ISO 9594-8/ITU-T Rec. X.509 (2001) The Directory: Public-key and attribute certificate frameworks

4. Scot Cantor. "Shibboleth Architecture, Protocols and Profiles, Working Draft 02, 22 September 2004, see http://shibboleth.internet2.edu/

5. David F. Ferraiolo, Ravi Sandhu, Serban Gavrila, D. Richard Kuhn And Ramaswamy Chandramouli. "Proposed NIST Standard for Role-Based Access Control". ACM Transactions on Information and System Security, Vol. 4, No. 3, August 2001, Pages 224–274

6. Internet2 Middleware Architecture Committee for Education, Directory Working Group (MACE-Dir) "EduPerson Specification (200312)", December 2003. Available from http://www.nmi-edit.org/eduPerson/internet2-mace-dir-eduperson-200312.html

7. C. Ellison, B. Frantz, B. Lampson, R. Rivest, B. Thomas, T. Ylonen. "SPKI Certificate Theory". RFC 2693, Sept 1999.

8. "OASIS eXtensible Access Control Markup Language (XACML)" v2.0, 6 Dec 2004, available from http://www.oasis-open.org/committees/tc_home.php?wg_abbrev=xacml

9. O. Bandmann, M. Dam, and B. Sadighi Firozabadi."Constrained delegation". In Proceedings of the IEEE Symposium on Research in Security and Privacy, pages131-140, Oakland, CA, May 2002. IEEE Computer Society Press.

10. Paul Madsen. "WS-Trust: Interoperable Security for Web Services". June 2003. Available from http://webservices.xml.com/pub/a/ws/2003/06/24/ws-trust.html

11. Markus Lorch , Seth Proctor , Rebekah Lepro , Dennis Kafura , Sumit Shah. "First experiences using XACML for access control in distributed systems". Proceedings of the 2003 ACM workshop on XML security, October 31-31, 2003, Fairfax, Virginia

12. Wolfgang Hommel. "Using XACML for Privacy Control in SAML-based Identity Federations". In 9th IFIP TC-6 TC-11 Conference on Communications and Multimedia Security (CMS 2005), Springer, Salzburg, Austria, September 2005

13. Alfieri R, et al. VOMS: an authorization system for virtual organizations, 1st European across grids conference, Santiago de Compostela. 13-14 February 2003. Available from: http://grid-auth.infn.it/docs/VOMS-Santiago.pdf

14. Tom Barton, Jim Basney, Tim Freeman, Tom Scavo, Frank Siebenlist, Von Welch, Rachana Ananthakrishnan, Bill Baker, Kate Keahey. "Identity Federation and Attribute-based Authorization through the Globus Toolkit, Shibboleth, GridShib, and MyProxy". To be presented at NIST PKI Workshop, April 2006.

15. Dwaine Clarke, Jean-Emile Elien, Carl Ellison, Matt Fredette, Alexander Morcos, Ronald L. Rivest. "Certificate chain discovery in SPKI/SDSI". Journal of Computer Security, Issue: Volume 9, Number 4 / 2001, Pages: 285 - 322

16. Y. Elley, A. Anderson, S. Hanna, S. Mullan, R. Perlman and S. Proctor, "Building certificate paths: Forward vs. reverse". *Proceedings of the 2001 Network and Distributed System Security Symposium (NDSS'01)*, Internet Society, February 2001, pp. 153–160.

17. Ninghui Li, William H. Winsborough, John C. Mitchell. "Distributed credential chain discovery in trust management".Journal of Computer Security 11 (2003) pp 35–86

18. XACML v3.0 administration policy Working Draft 05 December 2005. http://www.oasis-open.org/committees/documents.php?wg abbrev=xacml.

19. Housley, R., Ford, W., Polk, W., and Solo, D., "Internet X.509 Public Key Infrastructure Certificate and Certificate Revocation List (CRL) Profile," RFC 3280, April 2002

20. David Chadwick. "Authorisation using Attributes from Multiple Authorities" in Proceedings of WET-ICE 2006, June 2006, Manchester, UK

Appendix 1: CVS Policy Schema

```
<?xml version="1.0" >
<xs:schema xmlns:xs="http://www.w3.org/2001/XMLSchema"
xmlns:permis="http://sec.cs.kent.ac.uk/permis" elementFormDefault="qualified"
attributeFormDefault="unqualified">
<xs:element name="CVSPolicy" type="permis:CVSPolicyType"/>
   <xs:complexType name="CVSPolicyType" >
          <xs:sequence>
                  <xs:element name="TrustedIssuers" type="permis:TrustedIssuersType" />
                  <xs:element name="AttributeHierarchies"
type="permis:AttributeHierarchiesType" />
                   <xs:element name="Domains" type="permis:DomainsType"/>
                  <xs:element name="AttributeAssignments"
type="permis:AttributeAssignmentsType" />
          </xs:sequence>
          <xs:attribute name="CVSPolicyID" use="required" type="xs:anyURI"/>
   </xs:complexType>
<!-- -->
<xs:complexType name="TrustedIssuersType">
          <xs:sequence>
          <xs:element name="TrustedIssuer" maxOccurs="unbounded"
type="permis:TrustedIssuerType"/>
          </xs:sequence>
</xs:complexType>
<!-- -->
<xs:complexType name="TrustedIssuerType">
          <xs:attribute name="TrustedIssuer" use="required" type="xs:anyURI"/>
  <!-- Only LDAP and HTTP URLs are currently allowed for issuers -->
          <xs:attribute name="TID" use="required" type="xs:ID"/
</xs:complexType>
<!-- -->
<xs:complexType name="AttributeHierachiesType">
          <xs:sequence>
          <xs:element name="AttributeHierarchy" maxOccurs="unbounded"
type="permis:AttributeHierarchyType" />
          </xs:sequence>
</xs:complexType>
<!-- -->
<xs:complexType name="AttributeHierachyType">
          <xs:sequence>
          <xs:element name="Superior" type="permis:SuperiorValueType"
maxOccurs="unbounded" >
          <xs:sequence>
          <xs:attribute name="AttributeOID" use="required" type="xs:anyURI"/
          <!-- Must be encoded according to SAML LDAP Profile e.g. urn:oid:1.2.3.4 -->
          <xs:attribute name="FriendlyName" use="required" type="xs:ID"/
</xs:complexType>
<!-- -->
 <xs:complexType name="SuperiorValueType">
          <xs:sequence>
```

```
            <xs:element name="Subordinate" type="permis:SubordinateValueType"
minOccurs="0" >
            <xs:sequence>
            <xs:attribute name="Value" use="required" type="xs:ID" / >
 </xs:complexType>
<!-- -->
<xs:complexType name="SubordinateValueType">
            <xs:attribute name="Value" use="required" type="xs:IDREF"/
</xs:complexType>
<!-- -->
<xs:complexType name="DomainsType">
            <xs:sequence>
        <xs:element name="Domain" maxOccurs="unbounded" type="permis:DomainType" />
            </xs:sequence>
</xs:complexType>
<!-- -->
  <xs:complexType name="DomainType">
            <xs:sequence>
            <xs:element name="RootNode" type="permis:RootNodeType"
maxOccurs="unbounded"
            <xs:sequence>
            <xs:attribute name="DomainID" use="required" type="xs:ID"/ </xs:complexType>
<!-- -->
 <xs:complexType name="RootNodeType">
            <xs:sequence>
            <!-- the excluded nodes must be immediately subordinate to the root node.
            Only LDAP and HTTP URLs are currently allowed for nodes -->
            <xs:element name="ExcludedNode" type=" xs:anyURI " minOccurs="0"
maxOccurs="unbounded"
            <xs:sequence>
            <xs:attribute name="Name" type="xs:anyURI" use="required"/>
</xs:complexType>
<!-- -->
<xs:complexType name="AttributeAssignmentsType">
            <xs:sequence>
            <xs:element name="AttributeAssignment" maxOccurs="unbounded"
type="permis:AttributeAssignmentType"/>
            </xs:sequence>
</xs:complexType>
<!-- -->
<xs:complexType name="AttributeAssignmentType" >
             <xs:sequence>
            <xs:element name="Attribute" type="permis:AttributeType" minOccurs="0"
maxOccurs="unbounded" />
             </xs:sequence>
            <xs:attribute name="AAID" use="required" type="xs:ID"/>
            <xs:attribute name="TI" use="required" type="xs:IDREF"/
            <xs:attribute name="DomainID" use="required" type="xs:IDREF"/>
            <xs:attribute name="DelegationDepth" use="optional"
type="xs:nonNegativeInteger"/>
  </xs:complexType>
<!-- -->
```

```
<xs:complexType name="AttributeType">
        <xs:sequence>
        <xs:element name="AttributeValue" type="permis:SubordinateValueType"
minOccurs="0" >
        <xs:sequence>
        <xs:attribute name="FriendlyName" use="optional" type="xs:IDREF"/
</xs:complexType>
<!-- -->
</xs:schema>
```

One-Round Protocol for Two-Party Verifier-Based Password-Authenticated Key Exchange*

Jeong Ok Kwon[1], Kouichi Sakurai[2], and Dong Hoon Lee[1]

[1] Graduate School of Information Security CIST, Korea University
Anam-dong Seongbuk-Gu, Seoul, 136-701 Korea
{pitapat, donghlee}@korea.ac.kr
[2] Department of Computer Science and Communication Engineering
Kyushu University, 6-10-1 Hakozaki, Higashi-ku, Fukuoka, 812-0053 Japan
sakurai@csce.kyushu-u.ac.jp

Abstract. Password-authenticated key exchange (PAKE) for two-party allows a client and a server communicating over a public network to share a session key using a human-memorable password *only*. PAKE protocols can be served as basic building blocks for constructing secure, complex, and higher-level protocols which were initially built upon the Transport Layer Security (TLS) protocol. In this paper, we propose a provably-secure *verifier-based* PAKE protocol well suited with the TLS protocol which requires only a single round. The protocol is secure against attacks using compromised server's password file and known-key attacks, and provides forward secrecy, which is analyzed in the ideal hash model. This scheme matches the most efficient verifier-based PAKE protocol among those found in the literature. It is the first provably-secure *one-round* protocol for verifier-based PAKE in the two-party setting.

1 Introduction

Password-authenticated key exchange. To communicate securely over an insecure public network it is essential that secret keys are exchanged securely. The shared secret key may be subsequently used to achieve some cryptographic goals such as confidentiality or data integrity. Password-authenticated key exchange (PAKE) protocols in the two-party setting are used to share a secret key between a client and a server using *only* a shared human-memorable password. These password-only methods have many merits in views of mobility and

---

* The first and third authors were supported by the MIC (Ministry of Information and Communication), Korea, under the ITRC (Information Technology Research Center) support program supervised by the IITA (Institute of Information Technology Assessment) and the second author was supported by grant for International Cooperative Research from National Institute of Information and Communication (Theme: "A Research on Scalable Information Security Infrastructure on Ubiquitous Networks"). This work was done while the first author visits in Kyushu University, Japan.

H. Leitold and E. Markatos (Eds.): CMS 2006, LNCS 4237, pp. 87–96, 2006.
© IFIP International Federation for Information Processing 2006

efficiency. Naturally, they are less expensive and more convenient than smart cards and other alternatives. This password-only method can also eliminate the requirement of a public key infrastructure (PKI). Due to the merits, protocols for PAKE can be used in several environments, especially in mobile networks.

In mobile networks session key exchange for the secure communication services has to be done efficiently using relatively small resources. One of the main efficiency issues in real applications over mobile networks is how to reduce the number of rounds, the computing time, and the size of the transmitted messages since clusters of mobiles have memory and processing constraints, and the mobile networks have limited bandwidth. Especially, the number of rounds is very important a factor in case that session keys have to be exchanged frequently.

The difficulty to design a scheme for PAKE comes from the usage of a password having low entropy. For a human to easily memorize a password, a password may have low-entropy (i.e., 4 or 8 characters such as a natural language phrase). These natural language phrases are weak because they are drawn from a relatively small dictionary. So they are susceptible to dictionary attacks, also known as password guessing attacks. The fundamental security goal of PAKE is security against dictionary attacks. Usually, dictionary attacks are classified into two classes. In *on-line dictionary attacks*, an adversary attempts to use a guessed password by participating in a key exchange protocol. If the protocol run is failed, the adversary newly starts the protocol with the server using another guessed password. This attack requires participation of the adversary. In *off-line dictionary attacks*, an adversary selects a password from a dictionary and verifies his guess in off-line manner. Since the adversary uses only recorded transcripts from a successful run of the protocol, no participation of the adversary is required. So such off-line attacks are undetectable.

On-line dictionary attacks are always possible, but the attacks do not become a serious threat because they can be easily detected and thwarted by counting access failures. That is, a failed guess can be detected by the server since one can count the number how many somebody terminates the protocol with failure. However, off-line dictionary attacks are more difficult to prevent. Even if there exist tiny amounts of redundancy information in flows of the scheme, then adversaries can mount an off-line dictionary attack by using the redundancy as a verifier for checking whether a guessed password is correct or not. The main security goal of schemes for PAKE is to restrict the adversary to on-line dictionary attacks only.

In addition to dictionary attacks, a fundamental security goal of PAKE is *key secrecy*. This security level means that no computationally bounded adversary should learn anything about the session keys shared between two honest parties by eavesdropping or sending messages of its choice to parties in the protocol. Other desirable security goals are as follows (formal definitions are given in Section 2). The importance of the following attributes depends on the real applications. *Forward Secrecy* means that even with the password of the users any adversary does not learn any information about session keys which are successfully established between honest parties without any interruption. A PAKE protocol is secure against *known-key attacks* if the following conditions hold: First, compromise of multiple session keys for sessions other than the one does not affect its key secrecy. This notion of security means that session keys are computationally independent from each other. A bit more formally, this security protects against "Denning-Sacco" attacks [10] involving compromise of multiple session keys (for sessions other than the one whose secrecy must be guaranteed). Next, an adversary cannot gain the ability to performing off-line dictionary attacks on the users' password from using the compromised session keys which are successfully established between honest users.

Two models for 2-pary PAKE. PAKE protocols for 2-party are classified into two models according to the sameness of knowledge used in authenticated-key exchange: *Symmetric model* in which a client and a server use the same knowledge related with a password to authenticate each other and establish a session key. In usually, the client and the server own a password in plaintext form. *Asymmetric* (or *verifier-based*) *model* in which a client and a server use the asymmetric knowledge related with a password to authenticate each other and establish a session key. In usually, the client memorizes a password, while a server stores an image (called a verifier) of the password under a one-way function instead of a plaintext version of the password.

Most previous 2-party PAKE protocols have been constructed in the *random oracle model*. The random oracle model is a security model, where we assume that a certain function is an "ideal" function. In the ideal hash model, we assume a hash function is a random function and in the ideal cipher model, we assume that a block cipher is a random permutation.

Many provably-secure PAKE protocols in the *symmetric model* have been suggested [5,9,20,15,16,12,6,7,3,1]. In [5], Bellare *et al.* provided a formal model for PAKE and proved the security of a protocol of [4] in the ideal cipher model. Boyko *et al.* presented PAKE protocols provably-secure in the ideal hash model [9,20]. Katz *et al.* [15,16] and Goldreich *et al.* [12] proposed PAKE protocols provably-secure in the standard model, independently. Bresson *et al.* [6] proved the security of *AuthA* which is a PAKE protocol considered for standardization by the IEEE P1363 standard working group, in the ideal hash model and the ideal cipher model. Also Bresson *et al.* reduced the number of ideal functions and proved the security of *AuthA* in the ideal hash model [7]. Recently, Abdalla *et al.* proposed PAKE protocols provably-secure in the ideal hash model [3,1]. *Verifier-based* PAKE protocols has been extensively studied in the last few years: A-EKE, B-SPEKE, SRP, GXY, SNAPI-X, AuthA, PAK-Z+, AMP, EPA, and VB-EKE [8,14,23,19,21,5,11,17,13,2].

Server compromise in symmetric model. In a protocol of symmetric model, the client and the server own a password. Hence the corruption of the server reveals the passwords themselves and an adversary that is able to access to the server's password file, can immediately masquerade as a legitimate client by using only the corrupted password without executing of any off-line dictionary attack. To better understand the damage of the server compromise in the symmetric model, consider the protocol [3] in the symmetric model suggested by Abdalla *et al.* in Figure 1. In this protocol, it easy to see that if the server compromise occurs, an adversary who can access the compromised passwords, can immediately masquerade as a legitimate client C to the server since the adversary knows password pw of client C.

Public information: G, g, p, M, N, H
Secret information: $pw \in Z_p$

Client C		Server S
$x \xleftarrow{R} Z_p \; ; X \leftarrow g^x$		$y \xleftarrow{R} Z_p \; ; Y \leftarrow g^x$
$X^\star \leftarrow X \cdot M^{pw}$	$\xrightarrow{X^\star} \quad \xleftarrow{Y^\star}$	$Y^\star \leftarrow Y \cdot N^{pw}$
$K_C \leftarrow (Y^\star / N^{pw})^x$		$K_S \leftarrow (X^\star / M^{pw})^y$
$SK_C \leftarrow H(C, S, X^\star, Y^\star, K_C)$		$SK_S \leftarrow H(C, S, X^\star, Y^\star, K_S)$

Fig. 1. A PAKE protocol in symmetric model

Server compromise in verifier-based model. PAKE protocols in verifier-based model are designed to limit the damage due to the server compromise. In a verifier-based protocol, the client owns a password, but the server owns a verifier of the password. Hence the corruption of the server just reveals the verifier not the password itself. Of course the server compromise still allows off-line dictionary attacks, but even if the password file is compromised, the attacker has to perform additional off-line dictionary attacks to find out the passwords of the clients. It will give the server system's administrator time to react and to inform its clients, which would reduce the damage of the corruption. Therefore, the main security goal of verifier-based PAKE protocols is to force an adversary who steals a password file from a server and wants to impersonate a client in the file, to perform an off-line dictionary attack on the password file. The difficulty of off-line dictionary attacks on the corrupted password file depends on the difficulty of finding the original password from the verifier.

1.1 Our Work in Relation to Prior Work

Two-party PAKE protocols can be served as basic building blocks for constructing secure, complex, and higher-level protocols which were initially built upon the Transport Layer Security (TLS) protocol [22]. In this paper we focus on designing a round-efficient verifier-based two-party PAKE protocol that can be used in the key exchange phase of the TLS protocol. In the TLS protocol, the key exchange protocol is executed right after the "hello" flows in which the first is from the client to the server, then the second is from the server to the client. To improve round-efficiency, in the paper we assume that parties can transmit messages simultaneously. Actually, in many common scenarios parties are able to transmit messages simultaneously. By taking advantage of the communication characteristics of the network it may be possible to design protocols with improved latency. This is the focus of the present work.

Recently, a provably-secure one-round PAKE protocol in symmetric model achieving the goal is proposed by Abdalla *et al.* [3] and its forward secrecy is proved by Abdalla *et al.* in [1]. Because the protocol in [3,1] is PAKE protocol in symmetric model in which, to achieve authenticated key exchange, it must be assumed that a client and a server own information related with a password in the same form. On the other hand, in order to immune to attacks using compromised server's password file, in verifier-based PAKE, a client and a server use each other asymmetric information for a password to achieve authenticated key exchange. So the simple and novel approach in [3,1] can not be directly applied to verifier-based PAKE because of the critical assumption for possessing of the symmetric information of passwords. We note that converting a PAKE protocol in symmetric model into a PAKE protocol in verifier-based model is not easy at all. Since a mechanism converting a PAKE protocol in symmetric model to verifier-based PAKE protocol must not reveal any redundancy information that adversaries can mount an off-line dictionary attack. To solve this problem, we use an additional multiplicative function where the multiplicative function used in [13,3,1] multiplies the protocol messages by a value which is made with a password.

Table 1. Comparisons of complexity with the related verifier-based protocols

Scheme/ Resource	Round	Modular exponentiations		Communication		Security	Assumption
		Client	Server	Client	Server		
B-SPEKE [14]	3	2	1	$|c| + |\tau|$	$2|c| + |p|$	-	-
SRP [23]	3	2	1	$|p|$	$2|p|$	-	-
PAK-Z+ [11]	3	$1 + E_{S.gen}$	$E_{S.ver}$	$|p| + |\sigma|$	$|p| + 3|l|$	KK&FS	Ideal hash
AMP [17]	4	1	3	$|p| + |l|$	$> |p| + |l|$	-	-
EPA [13]	2	1	2	$|p|$	$|p|$	FS	Ideal hash
VB-EKE [2]	2	1	4	$3|p|$	$|p|$	-	-
Our Scheme	1	2	1	$|p|$	$2|p|$	KK&FS	Ideal hash

$S = \{S.key, S.gen, S.ver\}$: a signature scheme, $E_{S.gen}$: the number of exponentiations in signing, $E_{S.ver}$: the number of exponentiations in verifying, $|\sigma|$: the length of a signature, $|p|$: the length of a prime p of Z_p^* , $|l|$: the length of an output of a hash function, $|c|$: the length of a symmetric encryption, $|\tau|$: the length of a message authentication code. An FS protocol is a forward-secure key exchange protocol and a KK protocol is a secure key exchange protocol against known-key attacks.

We compare the resources of our protocol with the protocols, B-SPEKE [14], SRP [23], AMP [17], and PAK-Z+ [11] submitted to the IEEE P1363.2 standard proposal for Password-Based Public Key Cryptographic Techniques, and recently proposed protocols, EPA and VB-EKE. Table 1 summarizes the comparisons of complexity and security, where communication cost is the total number of bits that a client and a server send during a protocol run. In the comparison of computation cost, we are applying pre-computation technique to the protocols to minimize on-line computational overhead. EPA requires the smallest exponentiations and communication cost on the client side, and the smallest rounds among the previously suggested protocols. However, EPA has a type of "challenge-response" mechanism (i.e., firstly, the client sends a challenge message to the server in the first round and then the server sends a respond message generated by using the client's challenge messages in the second round. Finally, the client can compute the session key after receiving the server's message in the second round), so it is no longer possible to swap the flows by employing the advantage of simultaneous message transmission. We explore the possibility of designing a protocol for verifier-based PAKE which can be implemented in only *one-round* (assuming simultaneous message transmission). Our protocol gives a novel method to make it possible that the client and the server send independently their messages for the key exchange in a single round since the parties can add authentication to messages regardless of other parties' messages. Thus the client can compute the session key after receiving the server's message in the first round. On the other hand, our protocol is only slightly less efficient from a computational perspective on the client side than EPA. The proposed protocol is the first provably-secure verifier-based two-party *one-round* PAKE protocol providing forward secrecy in the ideal hash model.

2   Security Model

We use the standard notion of security as defined in [5] and used extensively since then. This will be necessary for proving the security about our schemes in later sections. We fix nonempty sets $\mathcal{C}$ of potential clients and $\mathcal{S}$ of potential servers. We consider a password-authenticated verifier-based key exchange protocol in which two parties, a client $C \in \mathcal{C}$ and a server $S \in \mathcal{S}$ want to exchange a session key.

Initialization. A party P may have many instances of the protocol, which is either a client or a server. An instance of P is represented by an oracle P^s, for any $s \in \mathbb{N}$. Each client $C \in \mathcal{C}$ holds a password pw_C obtained at the start of the protocol using a password generation algorithm $\mathcal{PG}(1^\kappa)$ which on input a security parameter 1^κ outputs a password pw uniformly distributed in a password space Password of size $\mathcal{PW}$. Each server $S \in \mathcal{S}$ holds a vector (the so called a verifier) $pw_S = [f(pw_C)]_{C \in \mathcal{C}}$ with an entry for each client, where f is a one-way function. We assume the set $\mathcal{S}$ contains a single server.

Partnering. Let sid_C^s be the concatenation of all messages that oracle C^s has sent and received. For the concatenation the messages are ordered according to the kinds of owners, e.g., the first part is the client messages and the server's messages are concatenated to them. Let a partner identifier pid_C^s for instance C^s be a set of the identities of the parties with whom C^s intends to establish a session key. pid_C^s includes C^s itself. The oracles C^s and S^t are *partnered* if $\mathsf{pid}_C^s = \mathsf{pid}_S^t$ and $\mathsf{sid}_C^s = \mathsf{sid}_S^t$.

Queries. An adversary $\mathcal{A}$ is a probabilistic polynomial-time machine that controls all the communications and makes queries to any oracle. The queries that $\mathcal{A}$ can use are as follows:

- Send(P^s, M): This query is used to send a message M to instance P^s (this models active attacks on the part of the adversary). When P^s receives M, it responds according to the key exchange protocol.
- Execute(C^s, S^t): This query models passive attacks, where the adversary gets the instances of honest executions of the protocol by C^s and S^t. (Although the actions of the Execute query can be simulated via repeated Send oracle queries, this particular query is needed to distinguish between passive and active attacks in the definition of forward secrecy.)
- Reveal(P^s): This query models the adversary's ability to obtain session keys, i.e., this models *known-key attacks* in the real system. The adversary is given the session key for the specified instance.
- Corrupt(P): This models exposure of the long-term key held by P. The adversary is assumed to be able to obtain long-term keys of parties, but cannot control the behavior of these players directly (of course, once the adversary has asked a query Corrupt(P), the adversary may impersonate P in subsequent Send queries.) We restrict that on Corrupt(P) the adversary only can get the long-term key, but cannot obtain any internal data of P.
- Test(P^s): This query is used to define the advantage of an adversary. This query is allowed only once by an adversary $\mathcal{A}$, and only to *fresh* oracles, which is defined later. On this query a coin b is flipped. If b is 1, the session key sk_P^s held by P^s is returned. Otherwise, a string randomly drawn from a session key distribution is returned.

Freshness. We define a notion of *freshness* considering forward secrecy which means that an adversary does not learn any information about *previously* established session keys when making a Corrupt query. We say an oracle C^s is *fresh* if the following conditions hold:

- C^s has computed a session key $sk_C^s \neq$ NULL and neither C^s nor S^t have been asked for a Reveal query, where C^s and S^t are partnered.
- No Corrupt(P) for any $P \in \text{pid}_C^s$ has been asked by the adversary before a query of the form Send($C^s, *$).

PAKE Security. Consider a game between an adversary $\mathcal{A}$ and a set of oracles. $\mathcal{A}$ asks the above queries to the oracles in order to defeat the security of a protocol $\mathcal{P}$, and receives the responses. At some point during the game a Test query is asked to a fresh oracle, and the adversary may continue to make other queries. Finally the adversary outputs its guess b' for the bit b used by the Test oracle, and terminates. We define CG to be an event that $\mathcal{A}$ correctly guesses the bit b. The advantage of adversary $\mathcal{A}$ must be measured in terms of the security parameter k and is defined as $\text{Adv}_{\mathcal{P},\mathcal{A}}(k) = 2 \cdot \Pr[\text{CG}] - 1$. The advantage function is defined as $\text{Adv}_{\mathcal{P}}(k, T) = \max_{\mathcal{A}}\{\text{Adv}_{\mathcal{P},\mathcal{A}}(k)\}$, where $\mathcal{A}$ is any adversary with time complexity T which is polynomial in k.

Definition 1. We say a protocol $\mathcal{P}$ is a *secure password-authenticated key exchange scheme* if the following two properties are satisfied:

- Validity: if all oracles in a session are partnered, the session keys of all oracles are same.

- Key secrecy: $\mathsf{Adv}_\mathcal{P}(k,T)$ is bounded by $\frac{q_{se}}{\mathcal{PW}} + \epsilon(k)$, where $\epsilon(k)$ is negligible. q_{se} is the number of Send queries and $\mathcal{PW}$ is the size of the password space.

(1) We say a protocol $\mathcal{P}$ is a secure PAKE scheme if validity and privacy are satisfied when no Reveal and Corrupt queries are allowed.
(2) We say a protocol $\mathcal{P}$ is a secure PAKE-KK scheme if validity and key secrecy are satisfied when no Corrupt query is allowed.
(3) We say a protocol $\mathcal{P}$ is a secure PAKE-FS scheme if validity and key secrecy are satisfied when no Reveal query is allowed.
(4) We say a protocol $\mathcal{P}$ is a secure PAKE-KK&FS scheme if validity and key secrecy are satisfied.

3   One-Round Verifier-Based PAKE Protocol for Two-Party

We now present our protocol, $\mathcal{VB}$-$\mathcal{PAKE}$ with implicit authentication for verifier-based PAKE in the two-party setting. In this paper, we assume the parties can transmit messages simultaneously.

$\mathcal{VB}$-$\mathcal{PAKE}$ can be seen a version for verifier-based PAKE of the protocol [3,1] in symmetric model. To convert the protocol in [3,1] into the verifier-based protocol, $\mathcal{VB}$-$\mathcal{PAKE}$, secure against server compromise attacks, we use an additional verifier and ephemeral Diffie-Hellman key exchange. We can easily see that $\mathcal{VB}$-$\mathcal{PAKE}$ is secure against server compromise attacks. Even if the verifiers, v_1 and v_2, are revealed to an adversary, the adversary can not immediately masquerade as C to S without off-line dictionary attacks since the adversary does not know pw of C. Only the client C knowing pw can compute g_1^y from Z and sk_C. The description of $\mathcal{VB}$-$\mathcal{PAKE}$ follows:

Public information. A finite cyclic group $\mathbb{G}$ of order q in $\mathbb{Z}_p^*$. Two primes p, q such that $p = 2q + 1$, where p is a safe prime such that the CDH problem is hard to solve in $\mathbb{G}$. g_1 and g_2 are generators of $\mathbb{G}$ both having order q, where g_1 and g_2 must be generated so that their discrete logarithmic relation is unknown. Hash functions $\mathcal{H}_i$ from $\{0,1\}^*$ to $\{0,1\}^{l_i}$, for $i = \{0,1\}$.

Initialization. A client C obtains pw at the start of the protocol using the password generation algorithm $\mathcal{PG}(1^k)$. C sends $v_1 = g_1^{\mathcal{H}_0(C\|S\|pw)} \bmod p$ and $v_2 = g_2^{\mathcal{H}_0(C\|S\|pw)} \bmod p$ which are verifiers of the password to a server S over a secure channel. Upon receiving the verifiers, S stores them in a password file with an entry for C.

Round 1. C chooses a random number $x \in \mathbb{Z}_q^*$, computes $\mathsf{X} = g_1^x \cdot v_2 \bmod p$, and sends (C, X) to S. S selects random numbers $y, z \in \mathbb{Z}_q^*$, computes $\mathsf{Z} = g_1^z \cdot v_2 \bmod p$ and $\mathsf{Y} = g_1^y \cdot v_1^z \bmod p$, and sends $(S, \mathsf{Y}, \mathsf{Z})$ to C.

Key computation. Upon receiving $(S, \mathsf{Y}, \mathsf{Z})$, C computes $T = (\mathsf{Z}/v_2)^{\mathcal{H}_0(C\|S\|pw)} \bmod p$, $K_C = (\mathsf{Y}/T)^x \bmod p$ and the session key $sk_C = \mathcal{H}_1(C\|S\|\mathsf{sid}_C\|K_C)$, where $\mathsf{sid}_C = \mathsf{X}\|\mathsf{Y}\|\mathsf{Z}$. Upon receiving (C, X), S computes $K_S = (\mathsf{X}/v_2)^y \bmod p$ and the session key $sk_S = \mathcal{H}_1(C\|S\|\mathsf{sid}_S\|K_S)$, where $\mathsf{sid}_S = \mathsf{X}\|\mathsf{Y}\|\mathsf{Z}$.

SECURITY ANALYSIS. We now present that under the intractability assumption of the CDH problem the proposed protocol is a secure key exchange protocol against dictionary attacks and known-key attacks and provides forward secrecy.

Theorem 1. Assuming $\mathbb{G}$ satisfies the CDH assumption, $\mathcal{VB}\text{-}\mathcal{PAKE}$ is a secure PAKE-KK&FS scheme when $\mathcal{H}_1$ is modeled as a random oracle. Concretely,

$$\mathsf{Adv}_{\mathcal{VB}-\mathcal{PAKE}}^{\mathsf{pake\text{-}kk\&fs}}(k, T, q_{ex}, q_{se}, q_h) \leq 4q_h N_s \mathsf{Adv}_{\mathbb{G}, \mathcal{A}}^{\mathsf{cdh}}(T) + \frac{q_{se}}{\mathcal{PW}} + \frac{(q_{ex} + q_{se})^2}{q},$$

where T is the maximum total game time including an adversary's running time, and an adversary makes q_{ex} Execute queries, q_{se} Send queries, and q_h Hash queries to $\mathcal{H}_1$. N_s is the upper bound of the number of sessions that an adversary makes, and $\mathcal{PW}$ is the size of the password space.

Proof of Theorem 1. The detailed proof of this theorem appears in the full version of the paper [18].

4   Concluding Remarks

All previous provably-secure verifier-based PAKE protocols have been constructed in the ideal hash model [11,13]. In this paper, we have also proposed a provably-secure protocol in the ideal hash model. However, no provably-secure verifier-based PAKE scheme in the standard model has been proposed yet. The difficulty is dealing with a pre-shared password for secure key agreement. Designing an efficient verifier-based PAKE protocol which is probably-secure in the standard model is the subject of ongoing work.

References

1. M. Abdalla, E. Bresson, O. Chevassut, A. Essiari, B. M öller, and D. Pointcheval, *Provably Secure Password-Based Authentication in TLS*, In Proc. of ASIACCS'06, ACM Press, pages 35-45, ACM Press, 2006.
2. M. Abdalla, O. Chevassut, and D. Pointcheval. *One-time Verifier-based Encrypted Key Exchange*, In PKC '05, LNCS 3386, pages 47-64, Springer-Verlag, 2005.
3. M. Abdalla and D. Pointcheval. *Simple password-based encrypted key exchange protocols*, In Proc. of CT-RSA 2005, LNCS 3376, pages 191-208. Springer-Verlag, 2005.
4. S. Bellovin and M.merritt. *Encrypted Key Exchange: Password-Based Protocols Secure against Dictionary Attacks*, In Proc. of the Symposium on Security and Privacy, pages 72-84. IEEE Computer Society, 1992.
5. M. Bellare, D. Pointcheval and P. Rogaway. *Authenticated key exchange secure against dictionary attack* In Eurocrypt '00, LNCS 1807, pages 139-155, Springer-Verlag, 2000.
6. E. Bresson, O. Chevassut, and D. Pointcheval. *Security Proofs for an Efficient Password-Based Key Exchange*, In Proc. of the 10th ACM Conference on Computer and Communications Security, pages 241-250, ACM Press, 2003.

7. E. Bresson, O. Chevassut, and D. Pointcheval. *New Security Results on Encrypted Key Exchange*, In Proc. of PKC 04, LNCS 2947, pages 145-158, Springer-Verlag, 2004.

8. S. Bellovin and M. Merritt. *Augmented encrypted key exchange: a password-based protocol secure against dictionary attacks and password-file compromise*, ACM Conference on Computer and Communications Security, pages 244-250, 1993.

9. V. Boyko, P. MacKenzie, and S. Patel. *Provably Secure Password-Authenticated Key Exchange Using Diffie-Hellman*, In Proc. of EUROCRYPT '01, LNCS 1807, pages 156-171, Springer-Verlag, 2001.

10. D. Denning and G. M. Sacco. *Timestamps in Key Distribution Protocols*, Communications of the ACM 24(8), pages 533-536, 1981.

11. C. Gentry, P. MacKenzie, and Z. Ramzan. *PAK-Z+*, Contributions to IEEE P1363, August 2005. Available from `http://grouper.ieee.org/groups/1363/`.

12. O. Goldreich and Y. Lindell. *Session-Key Generation using Human Passwords Only*, In Proc. of CRYPTO '01, LNCS 2139, pages 408-432. Springer-Verlag, 2001.

13. Y. H. Hwang, D. H. Yum, and P. J. Lee. *EPA: An Efficient Password-Based Protocol for Authenticated Key Exchange*, In ACISP '03, LNCS 2727, pages 452-463, Springer-Verlag, 2003.

14. D. Jablon. *Extended password key exchange protocols immune to dictionary attack*, In Proc. of WETICE97 Workshop on Enterprise Security, 1997.

15. J. Katz, R. Ostrovsky, and M. Yung. *Efficient Password-Authenticated Key Exchange using Human-Memorable Passwords*, In Proc. of EUROCRYPT '01, LNCS 2045, pages 475-494, Springer-Verlag, 2001.

16. J. Katz, R. Ostrovsky, and M. Yung. *Forward secrecy in Password-only Key Exchange Protocols*, In Proc. of SCN '02, LNCS 2576, pages 29-44, Springer-Verlag, 2002.

17. T. Kwon. Ultimate Solution to Authentication via Memorable Password, Contributions to IEEE P1363, May 2000. Available from `http://grouper.ieee.org/groups/1363/`.

18. J. O. Kwon, K. Sakurai and D. H. Lee. Full version of this paper, Available at `http://cist.korea.ac.kr/~pitapat/VBTS200610.ps`.

19. T. Kwon and J. Song. *Secure agreement scheme for gxy via password authentication*, Electronics Letters, 35(11):892-893, 1999.

20. P. MacKenzie. *More Efficient Password Authenticated Key Exchange*, In Proc. of the RSA Data Security Conference, Cryptographer's Track (RSA CT '01), LNCS 2020, pages 361-377, Springer-Verlag, 2001.

21. P.MacKenzie and R.Swaminathan. *Secure network authentication with password identification*, Presented to IEEE P1363a, August 1999.

22. M. Steiner, P. Buhler, T. Eirich, and M. Waidner. *Secure Password-Based Cipher Suite for TLS*, ACM Transactions on Information and System Security (TISSEC) 4(2):134-157, 2001.

23. T. Wu. *Secure remote password protocol*, In Proc. of the ISOC NDSS Symposium, pages 99-111, 1998.

Enhanced CAPTCHAs: Using Animation to Tell Humans and Computers Apart

Elias Athanasopoulos and Spiros Antonatos

Institute of Computer Science, Foundation for Research and Technology Hellas,
P.O Box 1385 Heraklio, GR-711-10 Greece
{elathan, antonat}@ics.forth.gr

Abstract. Completely Automated Public Turing Test to tell Computers and Humans Apart (CAPTCHA) is a –rather– simple test that can be easily answered by a human but extremely difficult to be answered by computers. CAPTCHAs have been widely used for practical security reasons, like preventing automated registration in Web-based services. However, all deployed CAPTCHAs are based on the static identification of an object or text. All CAPTCHAs, from simple ones, like typing the distorted text, to advanced ones, like recognizing an object in an image, are vulnerable to the *Laundry* attack. An attacker may post the test to a malicious site and attract its visitors to solve the puzzle for her. This paper focuses on sealing CAPTCHAs against such attacks by adding a dimension not used so far: animation. Animated CAPTCHAs do not have a static answer, thus even when they are exposed to laundering, unsuspected visitors will provide answers that will be useless on the attacker's side.

Keywords: Web Security, CAPTCHA, Laundry attacks.

1 Introduction

CAPTCHAs[7] are challenge-response puzzles used to determine whether a user is human or not. There are several types of CAPTCHA tests, including distorted text, pictures of objects or even audio clips in case of impaired users. A simple example of a CAPTCHA test is shown in Figure 1. Users are requested to type the text displayed in the picture, "smwm" in our example. Most advanced examples, like the one in Figure 2, ask the user to identify an object, a person or an animal.

CAPTCHA tests are dynamically generated by computers, in contrast to the standard Turing test which is administered by a human. This characteristic allows them to be widely used for practical security reasons. Their applications span across multiple domains, from preventing worms and spam to online polls and search engines. The most common application of CAPTCHA tests is the prevention of automatic registration in Web-based services, like Web-based e-mail. E-mail provider sites include a CAPTCHA test as a step of the registration process to stop bots from subscribing and using their resources for spam distribution. Other applications involve online polls, where CAPTCHAs ensure that

H. Leitold and E. Markatos (Eds.): CMS 2006, LNCS 4237, pp. 97–108, 2006.

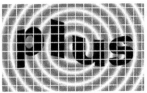

Fig. 1. An example **Fig. 2.** A sophisticated **Fig. 3.** A modern CAPTCHA, which is CAPTCHA, which is solved CAPTCHA, which adds solved if a user recognizes if a user identifies all the distortion to the image the word 'smwm' three animals

only humans vote or Web-blogs, where CAPTCHAs protect the blog from the massive insertion of garbage content by automated scripts. CAPTCHA tests can be circumvented in several ways. Advanced character recognition programs[13] can extract the text from simple tests like the one in Figure 1. However, tests used today are not that simple. By adding noisy backgrounds, colours and increasing the level of distortion, tests become resistant to character recognition programs. An example of how modern CAPTCHA tests look like is shown in Figure 3. Apparently, all CAPTCHA tests are vulnerable to laundry attacks. An attacker may post the test to her site and lure the visitors of this site to solve the test for her, e.g. by providing free access to content after the test is solved. Laundry attacks are independent of the complexity of current CAPTCHA tests. Their key property is that they use the intelligence of a human, thus any CAPTCHA tests in their current form are vulnerable to this attack.

In this paper, we present a novel technique for preventing laundry attacks for CAPTCHAs. The key idea behind our approach is that the answer of the CAPTCHA is embedded inside the test, animated to avoid static properties of current tests.. All current forms of CAPTCHAs follow the "type the answer" pattern, which dramatically helps laundering. We propose another form of CAPTCHA, where the answer is not static but floats around the test.

Specifically, our approach is a test where various objects are randomly moving inside the test. One of them is the correct answer and the user has to click on it to complete the test. Animation does not prevent the user to tell the attacker *what* is the answer, but prevents her from telling the attacker *where* is the answer.

2   Background

CAPTCHAs were originally developed by AltaVista. They were used to block or discourage the submission of URLs to their search engine. In 2002, Baird et al. developed PessimalPrint, a CAPTCHA that uses a model of document image degradations that approximates ten aspects of the physics of machine-printing and imaging of text. Their model included spatial sampling rate and error, affine spatial deformations, jitter, speckle, blurring, thresholding, and symbol size. BaffleText by PARC research uses non-English pronounceable character strings to defend against dictionary-driven attacks, and Gestalt-motivated image-masking degradations to defend against image restoration attacks.

Considerable research effort has been spent on breaking CAPTCHAs. Mori and Malik [13] have developed efficient methods based on shape context matching that can identify the word in an EZGimpy image with a success rate of 92%. Chellapilla et al.[10] have recently shown that computers are as good as or better than humans at single character recognition under all commonly used distortion and clutter scenarios used in todays CAPTCHAs. Poorly implemented CAPTCHAs can be also broken without using character recognition software but by exploiting session management weaknesses.

3   Animated CAPTCHAs

This Section aims at examining the threat model against current technologies used in the construction of CAPTCHAs and at describing our approach. To be fair, we give the attacker all benefits and we make no assumptions about the design of our approach. The key objectives of our approach is a) ease of deployment, we use industry standard technologies, such as Sun's Java Applets or Macromedia's Flash Movies, b) the test must be solvable by any user, at least as easy as current tests and finally c) robustness against attacks.

3.1   Laundry Attacks

Most services introduce a CAPTCHA to prevent automatic registration or ensure that a human is using them. Their main objective is to stop attackers from instrumenting their bots to automatically use the service for malicious purposes. An example is a Web e-mail service. In the absence of CAPTCHAs, the attacker could instruct her bots to register automatically to the service and start using the service in her way. For example, she can use the registered e-mail addresses to send spam. A CAPTCHA based on a static image is frequently used by large e-mail providers, such as Microsoft, Google and Yahoo, to ensure that the registration process was completed by a human and not a bot.

One way to defeat CAPTCHAs, based on a static image, such as the one in Figure 1, is by using sophisticated pattern matching. A bot can run special pattern recognition software that identifies the distorted word and eventually solve the test. However, the complexity of a static image CAPTCHA can be easily augmented and thus make the task of the pattern recognition program quite harder. Such an example can be seen in Figure 3. Although a human can easily identify that the "plus" word is displayed, the distortion in the picture increases exponentially the difficulty for a pattern recognition software. Thus, attackers are left with one solution to automatically solve a CAPTCHA: the laundry attack.

A laundry attack takes advantage of unsuspected users who will eventually solve a CAPTCHA in favor of the attacker, while they think that the CAPTCHA is solved for their own service. In more detail, consider an attacker who runs a popular Web site. Although it is out of the scope of this paper to explain how the malicious site will gain enough popularity, we can refer the reader to techniques[3].

Every visitor of the malicious site is lured to solve a CAPTCHA. However, this test is not generated by the site itself but it is actually the test of the victim service, for example the CAPTCHA of the registration phase of the Web e-mail service. No matter how difficult the test is, the answer now comes from a human and it is highly probable that it will be correct. The unsuspected user solves the test and the answer is "forwarded" to the victim service. In case the malicious site does not have enough popularity, it may present an aggressive behavior and periodically ask the user to solve a test, for example it can ask the user to solve one CAPTCHA per file download. In this way, the attacker has achieved to automatically solve CAPTCHAs independently of their difficulty, with a number of mistakes proportional to the number of mistakes ordinary users make and linearly to the number of visitors to her site.

The laundering of a CAPTCHA can be implemented by using the bots as intermediates. The malicious page that holds the "victim" puzzle contains a URL in the form "http://one_of_my_bots_IP/test.jpg". When an unsuspected user requests this URL, the bot sees the request and initiates the communication with the victim site, for example loads the registration page. Libraries like cURL [1] can be used to load pages from the command line offering full functionality similar to browsers, like cookies or redirections. The bot can also run a minimal HTML rendering engine and examine the loaded page, in order to spot the location of the CAPTCHA. Most sites have a constant name for the CAPTCHA image or even when they use dynamic names, their location inside the page is fixed or their names follow a specific pattern. After the test is located, it is then copied to "test.jpg" and served to the user. The user then answers the test to a form of the malicious site that has "http://one_of_my_bots_IP/submit.php" as action. The bot receives the answer and completes it to the loaded page. To the best of our knowledge, most services we tested allow multiple registrations per IP address, thus the attacker does not need to use all her bot power to perform the automated registrations, but a small fraction of it.

3.2   Animated CAPTCHAs to Prevent Laundry Attacks

Current forms of CAPTCHAs are subject to laundry attacks because of their static nature. They are pictures that contain the puzzle and the user has to complete the answer to a text field *outside* the puzzle. That is, the solution of the CAPTCHA is static and can be transfered between nodes of a malicious infrastructure (i.e. between a bot and a cooperative Web site which serves the laundry attack).

The first step we need to take is to transform a CAPTCHA test from a *static picture* to a *dynamic application*. That is, the answer must be completed *inside* the puzzle.

Consider a test where the user has to identify an object. The test is now a mini-application that contains both the image and the form where the user submits the answer. The form points directly to the service that originally creates it, for example the Web e-mail service, and the puzzle is immutable, e.g. the attacker cannot change its forms to redirect them to her bots. We see how we can achieve immutable tests in Section 4. The attacker now has to launder the advanced test. When an unsuspected user completes her answer inside the puzzle, her submission will go directly to the victim site and will fail. There are mainly two reasons for the failure. First, the solution was submitted with a different cookie or session ID as the request (some services use the PHP session ID to map the answer to the request). Second, the solution was submitted from a different IP address than the request.

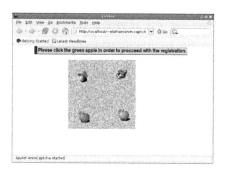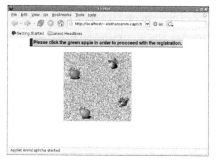

Fig. 4. An animated CAPTCHA, which we developed as a Java Applet

The attacker can circumvent this type of puzzles by posting a message like "Do not complete the answer inside the puzzle but to this text field". If the user follows the attacker's instructions (and we believe will do as she will have the incentives, e.g. access to the content) the same procedure as static puzzles can be followed. The second step is to eliminate the need for the user to type the answer and ask her to perform an action in the context of the application CAPTCHA. That is, the user is asked to *click* to the correct solution. The puzzle now contains multiple possible answers and the user has to click on the correct one. The click will submit the answer to the originating site. Again, the attacker may post a message like "Do not click on the test but complete this field where you would click". Although, the probability of a false answer has increased significantly (users may provide naive answers like "left" or "bottom") the attacker may assist users using Javascript snippets that can show the user the mouse coordinates.

The last step is to randomly animate the possible answers. While in the previous case answers remained static inside the puzzle, they now follow a random path. Even when the unsuspected user tells the attacker where she clicked, this information is useless at the attacker's side. The animation of the puzzle, which runs at the bot, is completely different than the one which runs at the user's side. Thus, an answer like "I clicked on x,y coordinates" is useless as in that location it can be any answer when the bot clicks. In Section 4 we will discuss in more detail randomness issues. Animated CAPTCHAs succeed to force attackers to try conventional methods of breaking the test, like brute-force or reverse engineering attacks (see Section 5) and not use the human intelligence of unsuspected users.

4   Implementation

As CAPTCHAs are mainly used in Web sites, in this Section we will focus on how we can implement animated CAPTCHAs for browsers. Our goal is to construct a CAPTCHA that cannot be modified by the attacker. Two possible implementation approaches are Flash movies and Java applets. Both of these types are widely available and can be found at most browsers. We implemented our prototype using Java technology.

We want to prevent the attacker from two actions: (a) the attacker must not be able to identify the correct answer via reverse engineering, (b) as answers trigger a communication with the originating service, the communication endpoint must not be circumvented.

4.1   Reverse Engineering

Although we can hide the source code for an animated CAPTCHA, it can be decompiled for source inspection and modification using freeware tools[5]. Assuming that the answer of the CAPTCHA is embedded in the application, a decompilation process could reveal it to the attacker. Our first option, towards this direction, is the use of code obfuscation freeware tools[9].

However, it is well known that a system can not base its security strength in obfuscation or secrecy[12]. But, in our case, we want to avoid fast and automated reverse engineering, which will not require human interaction.

In more detail, a Web site will generate an animated CAPTCHA during every registration process (or other activity, which must be verified that it is used by a human). Generation of an animated CAPTCHA means that the CAPTCHA will be compiled from a standard template, will be randomized by inserting random code and, finally, it will be implanted with the correct solution and the session ID of the host requesting the service. The correct solution is considered also a unique per CAPTCHA random token. In the Web site's side, we assume that there is a storage component to keep the mapping between correct tokens and session IDs. After the generation of the CAPTCHA, the resulting Java class file will be obfuscated. If the whole process of the generation and obfuscation of CAPTCHAs is considered a heavy job for the server to perform it on demand, it can use a pool of pre-generated CAPTCHAs (this pool can be maintained in low-traffic hours, in parallel with other maintenance procedures).

The above procedure guarantees that each animated CAPTCHA is a unique application. Each successful reverse engineering attempt, which should be also considered hard, must be triggered from a human and not by an automated program, since each CAPTCHA will have a different decompilation result. But, solving an animated CAPTCHA via human intervention is the definition of the CAPTCHA. The reverse engineering effort can not be re-used to solve automatically a collection of animated CAPTCHAs.

These ideas are already used in the case of polymorphic worms[14]. Code randomization has also been explored in various levels of software engineering and has been used to Computer Security[11].

4.2   Communication Circumvention

Our second goal is to protect the communication endpoint. Recall that the animated CAPTCHA is a Java applet, which embeds the token that maps in the solution and the session ID that maps to the host, which triggered the CAPTCHA. We need to prevent an attacker from locating the token or the session ID mechanically, since in that way she can create artificially a correct response to the service provider, and thus bypass the CAPTCHA.

   In order to deal with this issue the token of the correct answer and the session ID must be encrypted and located in a non-fixed place of the Java bytecode. We can achieve the latter using code randomization as explained in 4.1. The decryption key should also placed in a random location of the Java bytecode. An attacker can still reveal the decryption key, as well as spot the encrypted token and session ID through reverse engineering, but as we have already explained this process must be repeated for every CAPTCHA instance, since each CAPTCHA is a unique Java applet (in terms of bytecode). Thus, it is still impossible for an automated program to circumvent the communication channel without the human interaction.

5   Attacks Against Animated CAPTCHAs

With Laundry attacks eliminated, as it was described in Section 3.2, the malicious user will try to attack on the animated CAPTCHA itself. In this Section, we analyze in detail how an animated CAPTCHA can cope with attacks focused in animated CAPTCHAs. Furthermore, we implemented an actual animated CAPTCHA as a Java Applet (Figure 4), which had four objects following a circular orbit as possible solutions. During the implementation we tuned up various parameters in order to make the animated CAPTCHA more resistible against the attacks we describe below.

5.1   Brute Force Attack

The attacker may try to instruct her bot to continuously click on the puzzle until a possible answer is clicked. In that case, the probability to click a correct answer is $1/|possible answers|$. The number of possible answers cannot be high enough due to space reasons inside the puzzle. By placing tens of possible answers inside a limited space, the user will get confused and eventually she will be discouraged. Assume a bot-power (BP) of one thousand compromised machines and an animated CAPTCHA with ten possible answers. The probability for one member of the bot of solving randomly the CAPTCHA, $P(a)$, is $1/10$. Assume, also, that the under attack site allows a maximum of five retries (R) per IP address per day and only one registration per IP. We can estimate the amount of puzzles the attacker may solve in one day: $P = BP*P(a)*R = 1000*1/10*5 = 500$. That is, the attacker may succeed to solve 500 puzzles per day, and thus complete automatically 500 registration processes, which is considered too high.

In order to cope with the brute force attack, we need to reduce the probability $P(a)$. We can easily transform the answer space of the puzzle from a discrete to a contiguous one. This can be done, by treating *every* click as an answer, including the clicks that reached the blank space, which surrounds the animation. Thus, the probability of clicking the correct answer now depends on two factors: an answer is found under the point where bot clicks and this answer is correct. Thus the probability to solve the puzzle by random clicking is now equal to the ratio r, where r is the surface area of one answer divided by the surface area of the whole puzzle area. The animated Java applet we developed depicted in Figure 4 occupies a surface of 480x480 pixels and each possible answer is an icon, which occupies a surface of 48x48 pixels. That is the ratio r is 0.01. Using the same parameters as before and $P(a) = 1/10*r = 0.001$, the attacker can automatically solve 5 puzzles a day. By tuning the ratio, we can achieve one to two orders of magnitude reduction on the number of puzzles that can be automatically solved and have a user-friendly puzzle at the same time.

Moreover, by combining more animated CAPTCHAs the probability $P(a)$ is reduced drammaticaly and not in a linear fashion. For example, the test may require the user to click a group of moving animals in a specific order based on their size. If the animated CAPTCHA has ten moving objects and three of them are animals, then the probability of clicking an animal is $P(a) = 0.001$. The probability of solving the test is the product of the three individual probabilities of clicking an animal. That is $P(a) = (0.001)^3$.

5.2   Remote Control Attack - Sweatshop Attack

The attacker may proceed with a manual installation on each bot, through display redirection techniques. Sweatshops are also an available option [6]. An attacker can hire employees from a sweatshop who will proceed to manual installation on the bots. Employees connect to each bot and redirect their display to a local machine. Specialized software can be used for display redirection. We experimented with VNC[8]. VNC is a lightweight display redirector that is ported to most operating systems and can be easily installed in any system. We also considered the built-in functionality of the X11 server and the remote assistance feature of Windows XP professional. However, the X11 environment can be only found in Unix systems but most bots are Windows XP systems[15]. Furthermore, more bots are likely to have Windows XP Home installed. We ignore the fact that when VNC or remote assistance run, their presence is noticeable to the actual owner of the bot.

A way to defend against the sweatshop attack is to enhance the animation, and thus increase the bandwidth needed for the display redirector software. For example, observe that the animated CAPTCHA depicted in Figure 4 has a constantly changing background (similar to the snow effect of a non-working television) and that a multicolour display is required, since the user needs to be able to distinguish the green from the red apple. That is, a display redirector, configured in monochrome mode and in low resolution, so as to reduce the needed bandwidth cannot be used. Unfortunately, this choice has a drawback for people

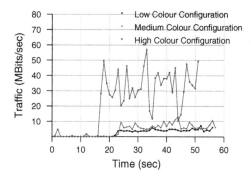

Fig. 5. VNC traffic for the various colour configurations. During the first 10 seconds we monitored the line without a VNC connection. The next 10 seconds we monitored the line with a VNC connection, but without running the CAPTCHA applet. During the rest of the period, the CAPTCHA applet runs remotely. Denote that we plot only the downstream traffic.

suffering of colour blindness, since they are not able to distinguish different colours[1]. However, W3C argues for the inaccessibility of all Visual CAPTCHAs for people with low vision[4].

In addition, if the answers inside the puzzle are animated slowly, the display redirection tools may be able to catch up with differences and display them correctly to the employees' machine.

We experimented on the delay introduced by VNC and bandwidth consumption for different display configuration between a hypothetical compromised machine and a hypothetical machine owned by an employer of a sweatshop. Both machines were interconnected in a LAN with 100 MBit/sec network connection. We collected three traces, using the Ethereal tool[2], for a colour display of 64 (Low Colour Configuration), 256 (Medium Colour Configuration) and 24-bit (High Colour Configuration), respectively. In each experiment we enabled the compression of the transmitted data, supported by VNC. Although, VNC supports an even lowest display configuration, with 8 colours, we did not collect a trace, since our CAPTCHA was impossible to be solved: the green apple (the correct answer) was displayed with a yellow colour.

The results are plotted in Figure 5. Observe that even at the lowest colour configuration, VNC introduces a network traffic closed to 6 MBit/sec in order to display the animated CAPTCHA. Denote, that the VNC connection is fired up after the first 10 seconds have elapsed, but the CAPTCHA is launched after the first 20 seconds have elapsed. That is, the enhanced animation of the CAPTCHA causes the VNC server to transmit more information in the VNC client.

Based on Figure 5, we understand that an attack based on the remote control of a compromised machine may succeed only if the compromised machine is equipped with a network connection closed to 6 MBit/sec or better. On the

---

[1] It is estimated that people suffering of colour blindness are almost 7% of all humans.

other hand, it is almost trivial to modify various properties in order to make the remote control of an animated CAPTCHA harder to be achieved.

Someone can argue that existing commodity network speeds may increase in the near future and thus remote control software will have the required bandwidth to transmit a complex animation. However, no matter the speed, someone can create a high quality multimedia CAPTCHA, including motion and sound information dependent with the solution, which will need enormous bandwidth in order to be solved using a remote desktop application. Apparently, a multimedia CAPTCHA that combines high quality motion, high quality sound and a possible sequence of logic actions, is out of the context of this paper, but is subject for our future work.

Table 1. A list of properties of an animated CAPTCHA, which can be easily tuned in order to make the CAPTCHA more resistible in possible attacks

	Property	Attack	Action	Result
1	Colourful answer	Remote Control	↑	Bandwidth consumption ↑
2	Animated background	Remote Control	Enhance	Bandwidth consumption ↑
3	Dimension of background	Brute Force	↑	$P(a) \downarrow$
4	Dimension of answers	Brute Force	↑	$P(a) \downarrow$
5	Background is an answer	Brute Force	+	$P(a) \downarrow$
6	Frame delay	Remote Control	↓	Bandwidth consumption ↑
7	Random orbit	Remote Control	+	Remote user difficult to adapt
8	Random frame delay	Remote Control	+	Remote user difficult to adapt
9	Code obfuscation	Laundry/RE	+	Reverse engineering effort↑
10	Code randomization	Laundry/RE	+	Reverse engineering effort↑
11	On the fly compilation	Laundry/RE	+	Reverse engineering effort↑

In Table 1 we summarize various parameters that someone can modify and make an animated CAPTCHA resistible in the attacks we presented, namely the basic Laundry, the Reverse Engineering, the Brute Force and the Remote Control/Sweatshop attack. Denote that the symbol ↑ means 'increasing', the symbol ↓ means 'reduced' and the symbol + means 'adding'. For example, consider Property 4, which is translated as: "Increasing (↑) the dimension of answers, during a Brute Force attack, the probability of a random guess $P(a)$ is reduced (↓)".

6   Conclusion and Future Work

In this paper, we investigated the state of the art of possible attacks against CAPTCHAs; puzzles that try to distinguish a human from a computer program, used mainly to prevent a service to malicious programs, such as bots. We introduced a new form of a CAPTCHA, which is based on animation. We argued that animated CAPTCHAs can resist to modern attacks, like laundering, and common attacks, such as brute-force or reverse engineering. In regards to well organized attacks via sweatshops, we measured the traffic required by a popular remote desktop software to run our animated CAPTCHA prototype remotely and showed that commodity user equipment is not sufficient. Finally, we suggested various enhancements, which will burden the task for an attacker to bypass an animated CAPTCHA, using either of the possible attacks and forcing her to manually solve the puzzles.

We believe that our enhanced with animation CAPTCHA technology can resist in sophisticated attacks better than standard CAPTCHAs based on static images with distorted text. However, we have not exposed our technology to the users and get their feedback, in order to understand the possible complexity which is introduced to ordinary Web surfers. Thus, we plan to evaluate animated CAPTCHAs against static CAPTCHAs and see how user-friendly our technology is, by performing experiments where real users must register to a service using a process that embeds static and animated CAPTCHAs.

In addition, part of our future work is multimedia CAPTCHAs. Puzzles that embed high quality motion, sound and a solution that is the result of a sequence of logic actions.

Acknowledgments. We would like to thank the anonymous reviewers for their valuable comments. We also thank Kostas G. Anagnostakis (I2R) for his insightful comments. This work was supported in part by the project CyberScope, funded by the Greek General Secretariat for Research and Technology under the contract number PENED 03ED440, and by the FP6 project NoAH, funded by the European Union under the contract number 011923. Elias Athanasopoulos and Spiros Antonatos are also with the University of Crete.

References

1. cURL. http://curl.haxx.se/.
2. Ethereal. http://www.ethereal.com.
3. Google bombing. http://en.wikipedia.org/wiki/Google_bomb.
4. Inaccessibility of CAPTCHA, Alternatives to Visual Turing Tests on the Web. http://www.w3.org/TR/turingtest/.
5. JCavaJ Java Decompiler. http://www.bysoft.se/sureshot/jcavaj/index.html.
6. Sweatshop. http://en.wikipedia.org/wiki/Sweatshop.
7. The CAPTCHA Project. http://www.captcha.net/.
8. VNC. http://www.realvnc.com.
9. yGuard. http://www.yworks.com/en/products_yguard_about.htm.
10. Kumar Chellapilla, Kevin Larson, Patrice Simard, and Mary Czerwinski. Computers beat humans at single character recognition in reading based human interaction proofs (hips). In *Second Conference on Email and Anti-Spam (CEAS)*, 2005.
11. Gaurav S. Kc, Angelos D. Keromytis, and Vassilis Prevelakis. Countering code-injection attacks with instruction-set randomization. In *CCS '03: Proceedings of the 10th ACM conference on Computer and communications security*, pages 272–280, New York, NY, USA, 2003. ACM Press.

12. A. Kerckhoffs. La cryptographie militaire. In *Journal des Sciences Militaires, 9 Jan 1883, pp. 5-38*. http://www.petitcolas.net/fabien/kerckhoffs/.
13. G. Mori and J. Malik. Recognizing objects in adversarial clutter – breaking a visual captcha. In *Conf. Computer Vision and Pattern Recognition, Madison, USA*, June 2003.
14. Peter Szoer and Peter Ferrie. Hunting for metamorphic. In *Virus Bulletin Conference*, September 2001.
15. The Honeynet Project Whitepapers. Know your enemy: Tracking botnets, March 2005. http://www.honeynet.org/papers/bots/.

Perturbing and Protecting a Traceable Block Cipher

Julien Bringer, Hervé Chabanne, and Emmanuelle Dottax

Sagem Défense Sécurité

Abstract. At the Asiacrypt 2003 conference, Billet and Gilbert introduce a block cipher, which, to quote them, has the following paradoxical property: it is computationally easy to derive many equivalent distinct descriptions of the same instance of the block cipher; but it is computationally difficult, given one or even many of them, to recover the so-called meta-key from which they were derived, or to find any additional equivalent description, or more generally to forge any new untraceable description of the same instance of the block cipher. They exploit this property to introduce the first traceable block cipher.

Their construction relies on the Isomorphism of Polynomials (IP) problem. At Eurocrypt 2006, Faugère and Perret show how to break this scheme by algebraic attack. We here strengthen the original traceable block cipher against this attack by concealing the underlying IP problems. Our modification is such that our description of the block cipher now does not give the expected results all the time and parallel executions are used to obtain the correct value.

Keywords: Traitor tracing, Isomorphism of Polynomials (IP) problem.

1 Introduction

Traitor tracing was first introduced by B. Chor, A. Fiat and M. Naor [4]. This concept helps to fight against illegal distribution of cryptographic keys. Namely, in a system, each legitimate user comes with some keys. We suppose that a hacker can somehow have access to them, maybe because some legitimate users are traitors. These keys can then be duplicated or new keys can be created by a pirate computed from legitimate ones. Traitor tracing enables an authority to identify one or all of the users in possession of the keys at the origin of the pirated ones.

Often traitor tracing is employed in a broadcast network. An encrypted signal is broadcasted and each legitimate user has the keys needed to decrypt it.

Today, many traitor tracing schemes are based on some key distribution and management techniques; the distribution of the keys is dependent on some combinatorial construction. A novelty comes in 1999 with D. Boneh and M. Franklin [3] (see also [12]) where public key cryptosystems are considered.

At the Asiacrypt 2003 conference, Billet and Gilbert [2] propose a traitor tracing scheme taking place at a different level as the block cipher which allows

H. Leitold and E. Markatos (Eds.): CMS 2006, LNCS 4237, pp. 109–119, 2006.

the decryption of the signal, also permits the traitor tracing functionality. To this aim, a block cipher which has many descriptions is introduced. All descriptions give – of course – the same result. Their idea relies on the Isomorphism of Polynomials (IP) trapdoor [14], based on algebraic problems for multivariate polynomials over finite fields. It was supposed that from one or many descriptions of this block cipher it is not possible to create new ones both allowing to decrypt the broadcasted signal and preventing the authority to trace back pirates. However, recently, Faugère and Perret [10] have presented a new algorithm for solving IP-like instances and have achieved to solve a challenge in [2]: namely, they break the instance of the scheme proposed by Billet and Gilbert.

We think it is worth trying to repair this scheme. Indeed, whereas most traditional traitor tracing schemes are combinatorial and bring about large overheads of encrypted data, this one – being non-combinatorial – avoids these large overheads. Furthermore, due to its symmetric-key-based nature it supports large numbers of users and is quite insensitive to the maximum number of tolerated traitors. Following the internal modifications of the Matsumoto-Imai cryptosystem from Ding [6], we add perturbations to Billet and Gilbert's traceable block cipher. Doing so, we want to protect the trapdoors from direct algebraic attacks (as for instance the recent algorithms of [8] and [10]), i.e. we want to alter the formal description of each round which forms the block cipher. However, here, we must still keep the traceable property with regard to the original block cipher. To manage this constraint, the perturbations are chosen in a particular way and we run in parallel, for each round, multiple descriptions of this round. None of them always gives the right result but we can show that a majority of these descriptions actually does, leading us to the expected value.

The paper is organized as follows. In Sect. 2, we recall a description of the traceable block cipher given by Billet and Gilbert. In Sect. 3, we give the principles of our modification of this traceable block cipher. In Sect. 4, we introduce the polynomials and techniques we use to fulfil our goal. In Sect. 5, we give practical implementations of our ideas. Starting from the examples given in [2], we describe their modified versions. We also show how to trace back pirates with our modified traceable block cipher. In Sect. 6, the security of the proposed scheme is analysed. Section 7 concludes.

2   A Traceable Block Cipher

The traceable block cipher of Billet and Gilbert is made of a succession of rounds. Each round is given by a system of equations in a finite field $\mathbb{F}$. The authority possesses a meta-key which allows it to compute the secret representations of the block cipher. The public representations consist of the suitable systems of polynomials $G_{i,j}$.

The left part of Figure 1 illustrates the secret authority description. Each round is made of a non-linear part preceded and followed by a linear transformation.

The invertible linear transformations $L_{i,j}$ depend on user j, the same is true for the order in which non-linear parts occur in the block cipher. We call σ_j this permutation of the rounds. Thus, for user j, the system of polynomials, giving his public representation of the rounds, is uniquely determined by the linear parts of the round $L_{i,j}$ and σ_j. It is computed from the secret representation by the authority and lies in the right part of Figure 1. For user j, we denote them by $G_{1,j}, \ldots, G_{r,j}$.

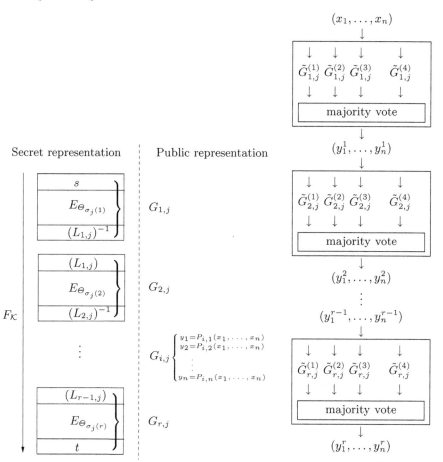

Fig. 1. A traceable block cipher   **Fig. 2.** New public representation

Here, r is the number of rounds, n stands for the number of variables, s, t and the $L_{i,j}$ are linear (s and t are shared by all users), the $E_{\Theta_{\sigma_j(i)}}$ are non-linear, and the polynomials $P_{i,1}, \ldots, P_{i,n}$ are homogeneous of degree d.

What made this block cipher traceable is the property that $E_{\Theta_{i_1}} \circ E_{\Theta_{i_2}} = E_{\Theta_{i_2}} \circ E_{\Theta_{i_1}}$, i.e. the non-linear parts commute, always leading to the same function $F_{\mathcal{K}} = t \circ E_{\Theta_{\sigma_j(r)}} \circ \cdots \circ E_{\Theta_{\sigma_j(1)}} \circ s$ independently of the order σ_j in which the rounds are given. The permutation σ_j on the order of the rounds is unique for each user and allows the authority to recover him. More precisely, to this aim of finding a user from his block cipher description, first, the authority computes in turn, for each $i \in \{1, \ldots, r\}$,

$$G_{1,j} \circ s^{-1} \circ E_{\Theta_i}{}^{-1}, \tag{1}$$

guessing the right value i by testing the simplicity of the result, i.e. by estimating the degree and the number of monomials. When $\sigma_j(1)$ has been found, the authority continues its procedure with $G_{2,j} \circ G_{1,j} \circ s^{-1} \circ E_{\Theta_{\sigma_j(1)}}{}^{-1} \circ E_{\Theta_i}{}^{-1}$, for $i \neq \sigma_j(1)$, trying to find back $\sigma_j(2)$, and so on, until the permutation σ_j is entirely recovered, see [2] for details.

Remark 1. *As pointed out by [1], another way to efficiently check the linearity of $G_{1,j} \circ s^{-1} \circ E_{\Theta_i}{}^{-1}$ is to search correlations between differential characterictics of the input and differential characteristics of the output.*

As explained in [2], the security of this scheme relies on the IP problem. In particular, an attacker, which is able to retrieve the polynomial $E_{\Theta_{\sigma_i(j)}}$ from the public representation $G_{i,j} = L_{i,j}^{-1} \circ E_{\Theta_{\sigma_i(j)}} \circ L_{i-1,j}$, could construct an untraceable description of $F_{\mathcal{K}}$. However, while the IP problem is considered as a hard problem in general, a new algorithm for solving instances of this problem is introduced in [10]: it allows, for some random or specific instances, via a fast Gröbner bases algorithm, to recover the secret isomorphisms from the knowledge of the public and the inner polynomials in an efficient way. For instance, they succeed in solving the first challenge of [2] in less than one second.

Our modifications of the scheme, which are introduced in the following, are then motivated by the fact that they need the formal description of the public polynomial to run the algorithm.

3   Our Protection in a Nutshell

We write $\tilde{0}$ for a polynomial which often vanishes and $\tilde{P} = P + \tilde{0}$. By the way, $\tilde{S}$ stands for a system S of equations where some substitutions are made, replacing some polynomials P by $\tilde{P}$.

Example 1. *Over $GF(q)[X]$, we have $\tilde{0} = X^{q-1} - 1$.*

Our idea is to simply replace $G_{i,j}$ by $\widetilde{G_{i,j}}$, for $i = 1, \ldots, r$. This way, the IP problem structure of each round is made less accessible to an attacker.

The construction where only one description of a round is modified is mainly given for pedagogical purpose and as an introduction to Sect. 5.2. Actually, it leads to wrong results.

In order to have a function which gives us always the correct result, we have to modify several instances of the block cipher. More precisely, we replace the system $G_{i,j}$ by 4 concurrent systems $\widetilde{G_{i,j}}$ where we can prove that two of them lead to what is expected. A majority vote allows to decide which result we have to retain. Note that this protection of one round can be seen as a protection of one IP-like instance, and this way, it could be applied to some other cryptographic schemes based upon IP.

4   Parasitizing the System with $\tilde{0}$-Polynomials

Example 1 is not sufficient because it does not allow enough diversity to stay hidden from an attacker. In this section, we introduce new $\tilde{0}$-polynomials to this aim. We proceed following two steps.

First, we introduce a well-known class of polynomials, the q_0-polynomials. With them, we are able to compute polynomials which vanish on a predetermined set of points. However, as q_0-polynomials are univariate and strongly related to vector spaces, next, we have to compose them with random multivariate polynomials.

4.1   Linearized Polynomials [13]

Definition 1. *For q_0 a power of 2 such that $q_0 \mid q$, a q_0-polynomial over $\mathbb{F} = GF(q)$ is a polynomial of the form $L(X) = \sum_{i=0}^{e} a_i X^{q_0^i}$, with $e \in \mathbb{N}$ and $(a_0, \ldots, a_e) \in \mathbb{F}^{e+1}$.*

Note that a q_0-polynomial L of degree q_0^e has at most $e + 1$ terms and a great number of roots in its splitting field. Indeed, if $a_0 \neq 0$, we see that L has only simple roots, so it has q_0^e zeroes in $\mathbb{F}$.

Example 2. *Let $\mathrm{Tr} : x \mapsto \sum_{i=0}^{15} x^{2^i}$ be the trace of $GF(2^{16})$ over $GF(2)$ and $\alpha \in GF(2^{16})$, then $L = Tr(\alpha.X)$ is a 2-polynomial with 16 terms and 2^{15} roots over $GF(2^{16})$.*

Proposition 1. *The set of a q_0-polynomial roots is a linear subspace of its splitting field, i.e. $L(X) = \sum_{i=0}^{e} a_i X^{q_0^i} = \prod_{\alpha \in V} (X - \alpha)^{\kappa}$ for V a linear subspace and some $\kappa \geq 1$. In fact, for a q_0-polynomial with simple roots, $\kappa = 1$.*

To count the number of q_0-polynomials with q_0^e roots of order 1, it suffices to count the number of $GF(q_0)$-subspaces of $GF(q)$ of dimension e:

Corollary 1. *For $q = q_0^m$, the number of q_0-polynomials with q_0^e roots of order 1 is equal to:*

$$G(q_0, m, e) = \frac{(q_0^m - 1) \cdots (q_0^{m-e+1} - 1)}{(q_0^e - 1) \cdots (q_0 - 1)}.$$

Due to the finite field structure, it is clear that a q_0-polynomial has at most 2^{m-1} roots, so, if we want to construct $\tilde{0}$-polynomials with more roots, we need to multiply several q_0-polynomials together. But, there would be some intersection among the roots of different polynomials. Hence, to increase the number of roots more efficiently, we can combine some affine q_0-polynomials which are the relevant construction of q_0-polynomials with an affine set of roots.

Definition 2. *For q_0 a power of 2 such that $q_0 \mid q$, an affine q_0-polynomial over $\mathbb{F} = GF(q)$ is a polynomial of the form $A(X) = L(X) - \alpha$ where $\alpha \in \mathbb{F}$ and L is a q_0-polynomial.*

4.2   Multivariate Lifting

In order to transform a q_0-polynomial into a multivariate polynomial, we compose it naturally with a multivariate polynomial.

Let Q be an affine q_0-polynomial over $GF(q^m)$ which equals zero over the subspace U of dimension e, we construct a multivariate version of Q by choosing a multivariate polynomial $f \in GF(q_0^m)[X_1, \ldots, X_{n_f}]$ and computing $Q_f = Q(f(X_1, \ldots, X_{n_f}))$. In our context, two conditions have to be considered :

1. the resulting polynomial must have at least the same proportion $\frac{1}{2^{m-e}}$ of roots as Q,
2. Q_f should not have a large number of terms.

Hence, we restrict the choice for f so as to respect the previous conditions. In practice, we take a random f with a small number of terms and we check if at least $1/2^{m-e}$ points of $GF(q_0^m)^{n_f}$ have an image following f in U. So the polynomial Q_f will have more than $2^{m.n_f}/2^{m-e}$ roots.

Example 3. *If $Q = \mathrm{Tr}_{GF(2^4)/GF(2)}(X)$, Q has 8 roots in $GF(2^4)$. Then the polynomial $f(X_1, X_2) = X_1 + X_1.X_2$ of $GF(2^4)[X_1, X_2]$ gives a polynomial Q_f with at least 32 roots in $GF(2^4)^2$.*

Eventually, this method allows to obtain a multivariate polynomial and also to randomize the construction by breaking its linear structure.

5   Some Practical Considerations

In Sect. 5 of [2], the authors provide two examples of a system for 10^6 users.

In the first one, the base field is $GF(2^{16})$ and there are 5 variables. The block cipher has 32 rounds and each equation is homogeneous of degree 4, hence each round has at most 350 monomials, and there is at most 11200 monomials for the whole system. We will refer to this example as the Case 1.

In the second one, which we call Case 2, the base field is $GF(2^9)$, there are 19 variables, the block cipher has 33 rounds and each equation is homogeneous of degree 3. So each round and the system have, respectively, at most 25270 and 833910 monomials.

5.1   Protecting One Round

In this section, we introduce a modified system leading to the correct result more than half time. In particular, we explain the interferences of our parasitic $\tilde{0}$ with the original public user representation; we show how we can choose some component H of $\tilde{0}$ to prevent an attacker to retrieve the original system.

Let $\tilde{0} = L(f(X_1, \ldots, X_{n_f}))H(X_1, \ldots, X_n)$ where

- L is a 2-polynomial with 2^{m-1} roots,
- f is a random polynomial of degree d_f in $2 \leq n_f \leq n$ variables and $t_f \geq 2$ terms such that $1/2$ of its values are roots of L,

- H is a random polynomial in n variables over $\mathbb{F}$ with t terms (H is more precisely described in the following).

Proposition 2. *The polynomial $\tilde{0}$ has about $N_1(m, t, t_f)$ terms and at least $1/2$ of roots where*

$$N(m, t, t_f) = m \times t \times t_f.$$

We add a parasitic $\tilde{0}$ to every equation of the round, taking the same 2-polynomial L for all equations of a given round but with different random polynomials H. This method allows the construction of a round function $\widetilde{G_{i,j}}$ that gives the correct result with a probability greater than $1/2$.

We introduce the polynomial H to generate enough monomials of degree d to avoid the capability of recovering P, a homogeneous multivariate polynomial of degree d, from the knowledge of $P + \tilde{0}$. In fact, starting from $P + \tilde{0}$, one can immediately compute the polynomial $\tilde{0}$ without its monomials of degree d, then knowing the form (i.e. designed as above) of $\tilde{0}$, one can try the two following ideas:

1. Guess the unknown monomials and their coefficients among all of the different possibilities, in order to obtain a polynomial with the same specific structure as $\tilde{0}$. There are $M_{n,d} = \binom{n+d-1}{d}$ monomials of degree d in n variables, so even if one guesses the number k of missing monomials, there would be $\binom{M_{n,d}}{k} q^k$ cases.

2. Analyse the terms of $P + \tilde{0}$ to guess the missing monomials, then, by deducing the generic form of H, try to find the missing coefficients by solving an overdefined system of equations, at least quadratic, in $t + l$ variables over $\mathbb{F}$ (where l is the number of variables coming from the unknown 2-polynomial of $\tilde{0}$ and from f). This kind of problem has been extensively studied these last years (see [5], [9] for example), and in general, one can not provide attacks in less than $q^{(t+l)/2}$, so we should consider t such that $q^t \geq 2^{160}$.

The choice of f and H is made in the following way: we choose f with at least one term of degree 1 in X_1 and if I is the set of $L(X_1)$ exponents, then we draw a polynomial H as

$$H(X_1, \ldots, X_n) = \sum_{i \in I \cap \{1, \ldots, 2^m - 1\}} h_i(X_1, \ldots, X_n) X_1^{2^m - i.d_f},$$

where the $h_i \in \mathbb{F}[X_1, \ldots, X_n]$ are homogeneous of degree $d - 1$. For each i, let t_i be the number of terms of h_i, then H has nearly $t = \sum_i t_i$ terms and the product $L(f(X_1, \ldots, X_{n_f}))H(X_1, \ldots, X_n)$ has at least t monomials of degree d. Hence, the number of monomials k which are masking the original polynomial P is greater than t, so a choice of t, such that $q^t \geq 2^{160}$ to avoid the second strategy above, allows also to thwart the first idea.

Furthermore, the number of choices for f and H is very large and so the amount of ways to interfere an equation is large enough.

Let us apply our strategy to the two practical examples of [2]:

- Case 1: We choose L, f, H such that $t_f = 2$ and $t = 10$, as described above. This implies $N(16, 10, 2) = 320$ terms more for each equation, and thus 1600 terms more for one round $\widetilde{G_{i,j}}$. This represents nearly 6 times the size of the original round.
- Case 2: For $t_f = 3$ and $t = 18$ such that $q^t \geq 2^{160}$, we have $N(9, 18, 3) = 486$ more terms for each equation. The resulting $\widetilde{G_{i,j}}$ has hence around 1.4 times the size of $G_{i,j}$.

Remark 2. *Roughly counting, there are more than* $\Lambda = G(2, m, m-1) \times \binom{d_f}{t_f - 1} \times 2^{m^{t+t_f}}$ *different ways to interfere an equation with such polynomials* $\tilde{0}$. *In case 1,* $\Lambda \geq 2^{208}$, *and in case 2,* $\Lambda \geq 2^{189}$.

5.2   Getting the Correct Value

For a given round $G_{i,j}$, we use four parallel modified descriptions $\widetilde{G_{i,j}}$ with correlated $\tilde{0}$-polynomials to recover the expected result.

To achieve this goal, we partition $\mathbb{F}$ and construct $\tilde{0}$-polynomials accordingly. As shown in Sect. 5.1, it is possible to cover more than half of $\mathbb{F}$. So, we partition $\mathbb{F}$ twice into two sets of the same size $\mathbb{F} = E_1 \cup \overline{E_1} = E_2 \cup \overline{E_2}$ and we construct $\tilde{0}_1$, $\overline{\tilde{0}}_1$, $\tilde{0}_2$ and $\overline{\tilde{0}}_2$ such that the polynomial $\tilde{0}_\kappa$ (resp. $\overline{\tilde{0}}_\kappa$) vanishes over E_κ (resp. over $\overline{E_\kappa}$), $\kappa = 1$ or 2.

Following this construction, for any input value, there are always two $\tilde{0}$-polynomials which vanish and so at least two descriptions $\widetilde{G_{i,j}}$ which give the expected result. Furthermore, as the constuction of an $\tilde{0}$-polynomial is partially random (see Sect. 5.1), the non-zero values of the two other $\tilde{0}$-polynomials look like random ones. Hence, with an overwhelming probability, the two other descriptions take two different results and so we can easily decide which value is correct according to a majority decision.

5.3   The Final Construction

Our new description of the entire public representation consists thus in modifying each round independently as described in Sect. 5.2. We obtain four parallel systems, with a majority vote at each level to decide which value has to be sent to the next round. See Fig. 2 for the resulting description.

Then, the size of this description according to the two practical examples of [2] is:

- Case 1: For the same choice of parameters as in Sec. 5.1, we have 1600 terms more for one round, i.e at most 1950 terms for each round. Thus, the final function (with 4 parallel systems of 32 rounds) contains around 22 times more terms than the original description.
- Case 2: Here, each equation contains at most 1816 monomials which leads to a final description with nearly 6 times the size of the original representation.

5.4   Tracing Procedure

Following [1] (see Remark 1), the authority can trace back pirates by a procedure relying only on the evaluation of rounds at given input, contrary to the way described in [2] which is based on compositions of polynomials. This method, via evaluations, is still compatible with our new description and can be used by the authority to trace back the traitors.

6   Security Analysis Overview

6.1   Multi Traitor Strategy

The traitor possibilities to use this new description in order to find another one or to collaborate are almost the same as in the original description [2]. Indeed, the information we add to the original system [2] is essentially random and gives no additional way to construct alternative untraceable polynomials.

6.2   Security of the Perturbed IP Problem

Our new procedure relies on a variation of the IP problem. The study of its actual security constitutes a challenging issue. As the formal description of a round is not given anymore, the attack from [10] does not hold. However, to this aim, the underlying IP problem (i.e. the original description of the round) must not be retrievable. We give here some hints of a proof. Given 4 parallel perturbations of a polynomial $P + \tilde{0}_1, \ldots, P + \tilde{0}_4$:

- One could consider each $P + \tilde{0}_i$ independently to recover P, but as previously discussed in Sect. 5.1, this is not practical due to the specific choice of the polynomials L, f and H.
- One can try to observe the majorities in order to exploit the correlations to recover P. The point is that, as each $\tilde{0}_i$ is constructed with independent randomized polynomials H, the $\tilde{0}_i$ are independent masking polynomials for P. They are correlated only on an evaluation point of view, so an opponent could exploit only the values of the 4 polynomials. Seeing the majorities, he can deduce as many roots of the $\tilde{0}_i$ as he can compute and then he could try to interpolate the polynomials $(\tilde{0}_1, \ldots, \tilde{0}_4)$. In this case, even if he guesses the polynomials $L \circ f$, the polynomials H (especially the number of terms t) are chosen to thwart such a strategy (see Sect. 5.1).
- One can think of using the differential cryptanalysis ideas from [11] where an attack against the perturbed HFE scheme proposed by Ding in [6] is described. First, it is interesting to notice that the perturbations here are more complex than those chosen by Ding in [6], and even in [7] whose goal is to prevent the attack [11]. Indeed our system is not only a noisy one but also an incorrect one. Secondly, the attack of [11] exploits the linearity of some differentials of the system (in HFE and in [6], the system is made of quadratic equations). In our case there are no such differentials, hence the attack seems not possible at all.

7   Conclusion

We here show how to repair Billet and Gilbert's traceable block cipher with respect to the recent attacks on IP instances [10]. In some cases, our additional protection leads to a 6-fold increase in keying material. As this traceable block cipher can be implemented in software on general purpose processors [2], we believe that this can be acceptable. Note that the technique we use to fix the problem, i.e. perturbing a system of multivariate poynomials by adding other polynomials whose values are zero with probability greater than 1/2, is quite general; one can think of its reuse in another context.

Acknowledgments

The authors thank gratefully Stéphanie Alt and Reynald Lercier for their comments on the tracing procedure. They also wish to thank the anonymous referees for their useful suggestions which improved the presentation of this paper.

References

1. Stéphanie Alt and Reynald Lercier, private communication.
2. Olivier Billet and Henri Gilbert, *A Traceable Block Cipher*, Advances in Cryptology – ASIACRYPT 2003 (C.S. Laih, Ed.), vol. 2894, 2003, pp. 331–346.
3. Dan Boneh, and Mathew Franklin, *An efficient public key traitor tracing scheme*, Advances in Cryptology – CRYPTO'99 (Michael J. Wiener, Ed.), vol. 1666, 1999, pp. 338–353.
4. Benny Chor, Amos Fiat, and Moni Naor, *Tracing Traitors*, Advances in Cryptology – CRYPTO'94 (Yvo Desmedt, Ed.), vol. 839, 1994, pp. 257–270.
5. Nicolas Courtois, A. Klimov, Jacques Patarin, and Adi Shamir, *Efficient Algorithms for Solving Overdefined Systems of Multivariate Polynomial Equations*, Advances in Cryptology – EUROCRYPT 2000, vol. 1807, 2000, p. 392 ff.
6. Jintai Ding, *A New Variant of the Matsumoto-Imai Cryptosystem through Perturbation*, Public Key Cryptography, 2004, pp. 305–318.
7. Jintai Ding and Jason E. Gower, *Inoculating Multivariate Schemes Against Differential Attacks*, PKC 2006 (M. Yung, Y. Dodis, A. Kiayiasan and T. Malkin, Eds.), vol. 3958, 2006, pp. 290–301.
8. Jintai Ding, Jason E. Gower and Dieter S. Schmidt, *Zhuang-Zi: A New Algorithm for Solving Multivariate Polynomial Equations over a Finite Field*, Cryptology ePrint Archive, Report 2006/038, 2006. http://eprint.iacr.org/.
9. Jean-Charles Faugère and Antoine Joux, *Algebraic cryptanalysis of Hidden Field Equation (HFE) cryptosystems using Gröbner bases*, Advances in Cryptology – CRYPTO'03 (Dan Boneh, Ed.), vol. 2729, 2003, pp. 44-60.
10. Jean-Charles Faugère and Ludovic Perret, *Polynomial Equivalence Problems: Algorithmic and Theoretical Aspects*, Advances in Cryptology – EUROCRYPT 2006 (Serge Vaudenay, Ed.), vol. 4004, 2006.
11. Pierre-Alain Fouque, Louis Granboulan and Jacques Stern, *Differential Cryptanalysis for Multivariate Schemes*, Advances in Cryptology, – EUROCRYPT 2005, (Ronald Cramer, Ed.), vol. 3494, 2005, pp. 341-353.

12. Aggelos Kiayias and Moti Yung, *Traitor Tracing with Constant Transmission Rate*, Advances in Cryptology – EUROCRYPT 2002 (Lars Knudsen, Ed.), vol. 2332, 2002, pp 450–465.
13. Rudolf Lidl and Harald Niederreiter, *Intoduction to finite fiels and their applications*, Cambridge University Press, 1986.
14. Jacques Patarin, Louis Goubin, and Nicolas Courtois, *Improved Algorithms for Isomorphisms of Polynomials*, Advances in Cryptology – EUROCRYPT'98 (Kaisa Nyberg, Ed.), vol. 1403, 1998, pp. 184–200.

A New Encryption and Hashing Scheme for the Security Architecture for Microprocessors

Jörg Platte, Raúl Durán Díaz, and Edwin Naroska

Institut für Roboterforschung, Abteilung Informationstechnik,
Universität Dortmund Germany
{joerg.platte, edwin.naroska}@udo.edu, raul.duran@uah.es

Abstract. In this paper we revisit *SAM*, a security architecture for microprocessors that provides memory encryption and memory verification using hash values, including a summary of its main features and an overview of other related architectures. We analyze the security of *SAM* architecture as originally proposed, pointing out some weaknesses in security and performance. To overcome them, we supply another hashing and protection schemes which strengthen the security and improve the performance of the first proposal. Finally, we present some experimental results comparing the old and new schemes.

1 Introduction

Protecting software is becoming more important for the future and therefore, efficient protection schemes are required. These schemes must provide a strong protection without requiring too many changes from the programmers point of view to be able to reuse existing code. Some processor extensions, like AEGIS [1] and *SAM* [2,3], have been built to provide a secure execution environment for programs. Using these extensions, a program can be protected to prevent program code and data based attacks as well as runtime attacks. Hence, they are suitable to implement efficient copy protection schemes which cannot be removed or bypassed. Additionally, they can be used to protect program code and data disclosure by using encryption of memory contents and they must be able to protect against software based attacks (e. g., administrator access) and hardware based attacks (e. g., bus sniffing).

Protecting data or program disclosure is important in case of remote execution of programs. For example, in GRID computing, programs can be executed on many different computers spread all over the world and the submitter of these programs may not trust all remote systems. Using a security extension, the GRID can be used even for sensitive simulation data or secret algorithms.

This paper provides a security analysis of *SAM's* security functions and proposes modifications to its hashing and encryption algorithms. Using this modified scheme the security can be enhanced and the hashing performance increased compared with the old scheme.

Section 2 provides a brief overview about the *SAM* security architecture. Other security architectures are presented in section 3. Section 4 analyzes the

H. Leitold and E. Markatos (Eds.): CMS 2006, LNCS 4237, pp. 120–129, 2006.

encryption and hashing functions and provides an optimized version. In sections 5 and 6 the simulation environment and the computed results are presented. Section 7 concludes this paper.

2 *SAM* Overview

SAM provides a secure execution environment for programs based on a standard processor design and a standard operating system. *SAM* aims at preventing tampering attempts as well as data and program disclosure.

The next paragraphs are providing a brief description of *SAM's* main attributes. A more detailed architectural description can be found in [2] and the design of the caches in [3].

The current *SAM* processor design is based on a SPARC V8 compatible CPU and was designed to be an optional extension. Hence, no secured bootstrapping or a persistent trusted operating system core is required to run *SAM* protected programs. Both protected and normal unprotected programs can be executed in a multitasking environment and small parts of the operating system are protected only while executing protected programs.

The processor core consists of an enhanced ALU supporting additional security instructions, an L1 data and instruction cache as well as an L2 cache. All data inside this core is trusted whereas all data outside is assumed to be untrusted. Hence, all data entering the L2 cache must be verified whereas all data written back to memory has to be protected against modifications and data disclosure. Data modifications are detectable by computing hashes for all protected cache lines. Prevention of data disclosure is achieved by transparent encryption of memory contents written to memory by the L2 cache. *SAM* uses the hashes both for memory protection and as a counter for a counter mode encryption algorithm to reduce the memory footprint. The hashing and encryption schemes are described in detail in section 4.

SAM uses a per process fixed virtual memory layout with two partly overlapping virtual address ranges. In the protected region all data is protected and verified by additional hash values and in the encrypted region, all data is additionally encrypted. All instructions located in protected memory regions (protected instructions) can access the decrypted memory contents in encrypted regions. Any other instruction can only access the encrypted data. Using virtual addresses simplifies paging of unused parts of the program or the hash tree.

SAM's caches have been suitably modified to provide these additional security functions. Additional security bits reflect the protection status for each cache line and dedicated security queues are used to hide additional latencies caused by verification and encryption/decryption. The memory write queues are computing hash values and encrypting cache lines to be written back to memory while the cache is able to process requests. A check queue contains all unverified cache lines and calculates hashes for these data in order to compare them with the ones in memory and detect memory based attacks.

All queues have a fixed size and therefore, stalls may occur when any of the queues is full. To prevent deadlocks, the L2-cache-bus arbiter monitors all queues and suppresses external cache accesses when the number of queue entries exceeds a given threshold. Figure 1 shows the relations between the L2 cache and the queues.

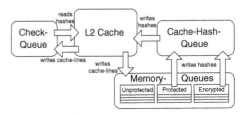

Fig. 1. Queue dependencies

In particular the check queue may exceed its threshold more likely, because each queue access may result in two additional queue entries, when requesting a hash results in a cache line replacement.

While *SAM* ensures that all internal data like variables, constants and functions are protected, all external data read from files or sockets could be potentially untrusted. Hence, the programmer has to check all external data by using suitable cryptographic protocols.

3   Related Work

Using cryptography to protect algorithms and data in a tamper resistant environment is not a new approach. Secure co-processors have been proposed which provide a tamper-sensing and tamper-responding secure environment. These processors can be implemented on smart cards (for example, [4]) or as a co-processor shown by [5] in a PC (for example, the IBM PCIXCC [6]). These co-processors provide a secure environment. But they are limited in terms of processor speed and memory and often, programs must be significantly modified to be suitable for this kind of co-processors. Therefore, they do not provide an easy-to-use and expandable secure environment.

Another more related approach is the AEGIS architecture [1], the successor of the XOM architecture [7]. Like *SAM*, AEGIS provides transparent data and instruction encryption, decryption and verification of memory contents.

In AEGIS, a program consists of unencrypted, encrypted and protected parts and the architecture provides secure transitions between these parts. Variables and functions can be assigned to these regions at compile time by the programmer. Hence, the programmer needs a profound knowledge about possible attacks to not leak secure data.

In order to prevent software based attacks, AEGIS requires a special bootstrapping mechanism to load a security kernel that has access to the page table and other sensitive information. Memory contents are protected by hash trees based on their physical address. This requires free pages at subsequent physical addresses and prevents paging of these data without reencrypting them.

Each time a new program is started AEGIS first computes a hash over all secured program related data. This hash is then used in conjunction with processor and operating system hashes to decrypt the program. The initial hashing of a program is a time consuming and complex task, which has been implemented by executing internal microcode instead of a direct hardware implementation.

In addition to the hash values, AEGIS suggests the usage of 32 bit counters to encrypt data resulting in approx. 6 % additional memory consumption. This potentially gives rise to more misses during memory operations. Depending on the memory access this counter can overflow resulting in a time consuming reencryption of all program related data with a new key. Longer counters can prevent this for most programs, but they consume more memory.

4   Hashing Scheme and Encryption

In this section, we first revisit the hashing scheme and data encryption proposed in [2,3], describing some weaknesses in security and performance. Then we propose some modifications on both the hashing and encryption schemes which improve the performance and strengthen the security.

4.1   Previous Hashing and Encryption Schemes

In general, a hash function h with round function f can be defined as follows:

$$H_0 = I_v, \quad H_i = f(x_i, H_{i-1}), \quad h(x) = H_m, \quad i = 1, 2, 3, \ldots, m \qquad (1)$$

where I_v stands for initialization vector and x_i are the m fixed-length blocks which comprise the input x (see, for example, [8, §9.4]).

Hashing scheme. First we explain how to hash one cache line as per paper [2]. A cache line C has 64 bytes (512 bits), further divided into four 16-byte (128-bit) blocks, for convenience. Each program is assigned a 128-bit key, k_s, at compile time.

A hash will be generated using AES as rounding function and a length of 128 bits for each block following the algorithm described in [9]. The value to be hashed is

$$C' = C \oplus (k_s || 0^{352} || V), \qquad (2)$$

where $||$ represents concatenation of bits and $\oplus$ is the XOR operator, k_s is the secret key, and V is the 32-bit cache line virtual address. This operation ensures that no data can be copied to another virtual address and that any two identical data hash to different values when located at different virtual addresses. The key k_s makes the hash value dependent on a secret value as well, and serves two purposes: firstly, it avoids possible exploitation of the hash value to extract information about the hashed contents, and secondly, several compilations of the same program will hash to different values, since k_s is randomly chosen for each compilation. Let $C' = (X_1 || X_2 || X_3 || X_4)$ be the value to hash, where each X_i is a 128-bit wide block, and let

$$f(x, H) = E_x(H) \oplus H \qquad (3)$$

be the rounding function, where $E_k(x)$ represents the application to x of the AES function with key k. Then the hash is computed as follows:

$$H_0 = 2^{128} - 1, \quad H_i = E_{X_i}(H_{i-1}) \oplus H_{i-1}, \quad H = H_5 = E_{H_4}(H_4) \oplus H_4, \quad i = 1, 2, 3, 4$$

Observe that this computation adds an extra step at the end, when compared with the general hashing equations (1). The authors claim in [9] that this step is necessary since, otherwise, "an attacker could take a hash without knowing the corresponding file (i.e., the value to be hashed), and use it to generate the hash of a file which is the original file with an appended bitstring of arbitrary content." This hash generation needs five applications of the AES function.

However, data integrity cannot be guaranteed using only the previous scheme, because replay attacks are still possible. For this reason, Merkle trees (see [10]) are used, whereby each hash cache line, consisting of four hashes, is in its turn hashed and stored in a sort of tree manner. The uppermost level is called the root hash and consists only of one hash value, which protects the last four hash values, and it is stored permanently inside the processor. See [2,3] for more details.

Data encryption. Each cache line can be further encrypted using AES with the secret key k_s in counter mode (see [11]). In this mode, an arbitrary value (the counter) is encrypted with the key k_s and XORed with the plain data to encrypt, in this case, the cache line. The hash value described above is used as a counter. However, each cache line consists of four blocks, and so one hash value does not suffice as a counter, since this would mean reusing the same counter for four different blocks. To avoid this problem, it was suggested in [2] to XOR the hash value with four arbitrary (but architecturally fixed) bit patterns, $R_1, \ldots, R_4$, which could thus supply four different counter values.

Performance evaluation. The obvious drawback of the described algorithm is that it is completely serial. Actually, the AES unit must used five times to compute the hash of one cache line. Therefore, the speed of one cache line hash computation can be only improved with a faster AES unit.

Security analysis. The definition of C' as per equation (2) presents the following undesirable property: two different cache lines, C_1, C_2 at virtual addresses V_1, and V_2, such that $C_1 \oplus V_1 = C_2 \oplus V_2$, will produce the same C' and, hence, the same hash value. But, then, this means that if C_1 and C_2 are to be encrypted, they will use the same counter, which is completely unsafe.

Besides, the key k_s serves two different purposes: it is used both to generate C' and to encrypt data. It would be advisable to avoid this, just in case unexpected cryptographic primitive interactions may arise.

Last, it would be very interesting to get rid of the four arbitrary bit patterns $R_1, \ldots, R_4$, if possible.

4.2   New Proposal for Hashing and Encryption

The purpose in this section is to describe our new proposal for both the hashing and the encryption schemes, which improves the speed of operation, while even increasing the level of security.

New hashing. The new hashing scheme has been suggested in [12, §2.4.4]. The idea is to replace the linear structure by a tree structure. This scheme, while not new, allows for a substantial speed-up in the evaluation of the hash function, which is now reduced to $O(\log m)$, where m is the number of blocks to hash.

Suppose, as above, that C is the cache line to hash. First of all we compute $C' = C \oplus r \oplus V$, where r stands for a random value generated at compile time, and V is the virtual address of the cache line. The reasons to use the values r and V are the same as in the old scheme. Note, however, that the computation of C' does not preclude the problem that two different cache lines located at different virtual addresses could receive the same hash value. This problem will be addressed below.

We will use the same rounding function described in equation (3). Assuming again that $C' = (X_1||X_2||X_3||X_4)$, where each X_i is a 128-bit wide block, the hash is computed as follows. First the following operations are performed in parallel:

$$H_1 = E_{X_1}(X_2) \oplus X_2, \qquad H_2 = E_{X_3}(X_4) \oplus X_4,$$

where $E_k(x)$ represents, as before, the application of the AES function to x with key k. When the previous step is over, the following computation is carried out:

$$H = E_{H_1}(H_2) \oplus H_2,$$

Remark that in this case only two serial applications of AES are needed, versus five applications in the old scheme; this means a speed-up of roughly[1] $5/2$. Last, observe that a final AES application is not needed (and thus can be saved), since the input has a fixed size and, therefore, the attack claimed by the authors in [9] cannot succeed in this particular case.

New encryption. In this new scheme, AES in counter mode will be used, as before, but in a slightly different manner. First of all, each program will receive at compile time a base encryption key k_b. The encryption will be performed now at the block level, using a different key to encrypt each block; this encryption key will be derived from the virtual address of the block to be encrypted, using k_b as a parameter. This "derived key" will be the actual encryption key for the block to encrypt.

More precisely, let $\mathcal{K}$ be the space of encryption keys, let $\mathcal{K}^\star$ be the space of base keys, and let $\mathcal{V}$ be the space of virtual addresses; then, given a block X, which belongs to cache line C and is located at virtual address V, the encryption function is

$$E_{g_{k_b}(V)}(H(C)) \oplus X. \tag{4}$$

In this equation, $g \colon \mathcal{K}^\star \times \mathcal{V} \to \mathcal{K}$ is a suitable transformation function, such that, for each value of the parameter $k_b \in K^\star$, $g_{k_b}(V)$ supplies a usable AES encryption key. This function should satisfy the following property: for any k_b, $k_b' \in \mathcal{K}^\star$, then there do not exist $V, V' \in \mathcal{V}, V \neq V'$, satisfying $g_{k_b}(V) = g_{k_b'}(V')$. In practice, such function exists since $|\mathcal{V}| \ll |\mathcal{K}|$, but then, of course, the base key space is smaller than the original, namely, $|\mathcal{K}^\star| < |\mathcal{K}|$.

Initially, the requirement of hardware simplicity compels us to use a simple g, such as XORing k_b and V. But, then, the system could be liable to a related-key attack, as described below.

---

[1] Some clock cycles are needed to initiate the second application of AES in the first step, so both operations are not strictly parallel.

Security analysis. The encryption scheme is based on the use of the counter mode. As stated in [11], it is required that a unique counter is used for each plain text block that is ever encrypted under a given key. There is no particular indication on the counting values, as long as they satisfy the uniqueness requirement. This makes it possible for us to use the hash values as counters, since the design of g and the protocol guarantee that they are unique for a given derived key. In fact, as it is easily checked, they only depend on the contents of the particular block to be encrypted.

Remark also that the new encryption method eliminates the need for the constants $R_1, \ldots, R_4$ described in section 4.1, since the encryption is now performed on a block basis. Besides, the random value r has nothing to do with the encryption key k_s of the old scheme, thus effectively decoupling both operations.

Finally, in order to keep an overall good performance, g should be evaluable in a short time frame (for example, one clock cycle).

4.3   Revision of Some Common Attacks

We will subsequently revisit some possible attacks.

Random attack. The opponent selects a random value and expects the change will remain undetected. If the hash function has the required random behavior, the probability of success is $1/2^n$, where n stands for the number of bits in the hash. In our case, this attack is not feasible, since we are using 128 bit hashes.

Birthday attack. In a group of at least 23 people, the probability that at least two people have a common birthday is greater than $1/2$: this is called the *birthday paradox*. This fact inspires the so-called birthday attack, applicable when an adversary tries to generate a collision. Remember that the hashes are stored in a Merkle tree fashion, all of them in plain text except for the root hash, which is kept encrypted. The attacker is then forced to face the problem of finding a preimage for any of the hashes, since a collision is of no use. Therefore, this attack is not applicable. Moreover, a random preimage attack on a 128-bit hash code requires 2^{128}, which can be considered unfeasible.

Related-key attack. In this attack, the enemy is allowed to observe the operation of a cipher under different but mathematically related keys. In our case, cache lines are liable to hash to the same value under certain conditions, as stated above. Therefore, it is advisable that the transformation function g used in equation (4) be selected so as to satisfy the necessary degree of randomness allowing the different "derived keys" to not disclose any mutual relationship.

5   Simulation Environment

This section briefly describes the simulation environment used to compute the results presented in section 6. The performance evaluation of different cache configurations is based on the SPEC benchmark suite. All benchmarks are executed in a virtual machine emulating a SPARC based computer with peripherals like hard disk, framebuffer and keyboard. This virtual machine is based on the free system emulator QEMU [13]. QEMU achieves a good performance by translating all instructions of the guest system to native assembler instructions of the host system. Hence, all timing and memory access information are lost. Therefore, QEMU has been extended to add special monitoring instructions during the translation step to reveal this lost information. They are used to log instruction fetches, read and write data and I/O accesses by the CPU and memory access by simulated peripheral devices as well as context switches, interrupts, and the current clock cycle to a trace file.

Table 1. Cache properties

Cache property	Value
L1 placement	direct mapped
L1 line size	32 bytes
L2 placement	LRU, 4-way-set
L2 line size	64 bytes

Bus	Width	Divisor
L1 $\leftrightarrow$ L2 cache	128 bit	2
to memory	64 bit	5
L2 cache $\leftrightarrow$ Queues	128 bit	2
to AES units	128 bit	2

Table 2. Cache configurations

Name	L1 size	L2 size	AES units	Check Queue entries
8-256	8k	256k	5	3
16-1024	16k	1024k	5	3
32-2048	32k	2048k	5	3

This trace file is then used as an input file for the *SAM* cache simulator. It simulates an L1 data and instruction cache as well as the L2 cache with all security related queues as described in section 2 to compute the number of simulated clock cycles for these operations. Instruction and data access is passed to the corresponding L1 cache and external device access is simulated by occupying the memory bus. One limitation of using a trace file is the missing feedback from the simulator to QEMU.

The cache simulator is fully configurable in terms of cache sizes, bus widths, number of queue entries and their thresholds, clock divisors to simulate different clock rates for buses and components like the queues or the caches, memory latencies or hashing algorithms. The L1 cache runs always with maximum clock speed and all other components are clocked with divisors based on this clock rate. Table 1 lists the basic configuration used for all simulations.

For all simulations, all data between the virtual addresses 0x70000000 and 0xf0000000 has been encrypted. The hash tree starts at address 0x1aaaaab0. A slightly modified Linux kernel has been used for the simulations. The kernel now starts to allocate memory for the heap starting at address 0x80000000 and all benchmarks are statically linked to the base address 0x70000000.

For each simulated benchmark 2^{32} instructions have been written to a trace file after skipping the first 2^{32} instructions, approximately, which correspond basically to the initialization routines of each benchmark. Using this trace file a set of different cache configurations has been simulated to obtain the overall number of simulated cache clock cycles needed for all cache operations. This set includes a configuration without security extensions which is further used as a reference for the speedup computation. The trace file does not contain any hash related data. During the simulation the cache simulator provides a random mapping of hash values to unused physical pages.

6   Simulation Results

Table 2 gives an overview about the configurations used for all simulations. The extension HT and HS are used for the tree and the sequential hashing algorithms, respectively.

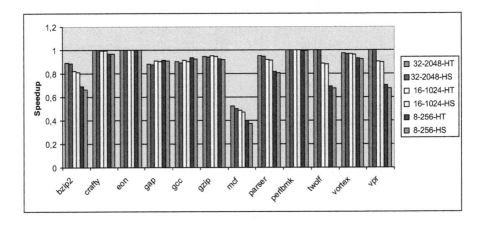

Fig. 2. Results for different cache sizes

Figure 2 compares both the sequential and the tree hashing algorithm for three different cache configurations. For nearly all configurations the speedup using the tree algorithm is higher than for the sequential algorithm. Also, the tree algorithm allows a more effective usage of the available AES units.

Using a larger cache does not always result in higher speedup as can be seen, for example, in the *gcc* benchmark[2]. Further investigation revealed that the number of cache line replacements (even for the cache configuration without security extensions) for those benchmarks is very similar for all three simulated cache configurations though the number of clock cycles is higher for the smaller caches. As a result, in this case even minor effects like different random mappings of hash values to physical pages (and therefore different sets) can distort this result.

---

[2] Remember, that the speedup for each benchmark has been computed in comparison with an equally configured cache without security extensions.

7   Conclusion

In this paper the cryptographic part of a processor security extension has been analyzed and optimized. The algorithm used for hash value generation has been optimized to provide a faster hardware implementation by parallelizing the algorithm and occupying more AES units at the same time. The presented benchmark results show that the presented new hashing algorithm further improves the good performance of the old scheme even without increasing the number of AES units.

The security analysis of the encryption scheme used by *SAM* does not prevent hash collisions for all cases. However, the proposed algorithm reduces the probability of collisions while slightly increasing the overall performance.

As a result, hash values can be used both for memory integrity verification and as a counter for a counter mode encryption scheme. Thus a significant saving in memory can be achieved compared to other architectures.

References

1. Suh, G.E.: AEGIS: A Single-Chip Secure Processor. PhD thesis, Massachusetts Institute of Technology (2005)
2. Platte, J., Naroska, E.: A combined hardware and software architecture for secure computing. In: CF '05: Proceedings of the 2nd conference on Computing frontiers, New York, NY, USA, ACM Press (2005) 280–288
3. Platte, J., Naroska, E., Grundmann, K.: A cache design for a security architecture for microprocessors (SAM). In Grass, W., Sick, B., Waldschmidt, K., eds.: Lecture Notes in Computer Science. Volume 3894. (2006) 435 – 449
4. Sun Microsystems: Java card security white paper. http://java.sun.com/products/javacard/JavaCardSecurityWhitePaper.pdf   (2001)
5. Yee, B.: Using secure coprocessors. PhD thesis, Carnegie Mellon University (1994)
6. Arnold, T.W., Van Doorn, L.P.: The IBM PCIXCC: A new cryptographic coprocessor for the IBM eServer. IBM Journal of Research and Development **48** (2004) 475–487
7. Lie, D., Thekkath, C.A., Mitchell, M., Lincoln, P., Boneh, D., Mitchell, J.C., Horowitz, M.: Architectural support for copy and tamper resistant software (2000)
8. Menezes, A., van Oorschot, P., Vanstone, S.: Handbook of Applied Cryptography. CRC Press, Inc., Boca Raton, FL (1997)
9. Cohen, B., Laurie, B.: AES-hash. http://csrc.nist.gov/CryptoToolkit/modes/proposedmodes/aes-hash/aeshash.pdf (2001)
10. Merkle, R.C.: Protocols for public key cryptosystems. In IEEE, ed.: IEEE Symposium on Security and Privacy, 1109 Spring Street, Suite 300, Silver Spring, MD 20910, USA, IEEE Computer Society Press (1980) 122–134
11. Dworkin, M.: Recommendation for Block Cipher Modes of Operation. Methods and Techniques. NIST. (2001)
12. Preneel, B.: Analysis and Design of Cryptographic Hash Functions. PhD thesis, Katholieke Universiteit Leuven (Belgium) (1993)
13. Bellard, F.: QEMU. http://fabrice.bellard.free.fr/qemu (2005)

Timed Release Cryptography from Bilinear Pairings Using Hash Chains

Konstantinos Chalkias and George Stephanides

Department of Applied Informatics, University of Macedonia, Thessaloniki, Greece
chalkias@java.uom.gr,
steph@uom.gr

Abstract. We propose a new Timed Release Cryptography (TRC) scheme which is based on bilinear pairings together with an S/Key-like procedure used for private key generation. Existing schemes for this task, such as time-lock puzzle approach, provide an approximate release time, dependent on the recipients' CPU speed and the beginning time of the decryption process. Additionally, some other server-based schemes do not provide scalability and anonymity because the server is actively involved in the encryption or the decryption. However, there are already protocols based on bilinear pairings that solve most of the problems referred. Our goal is to extend and combine the existing protocols with desirable properties in order to create a secure, fast and scalable TRC scheme applied to dependent or sequential events. For this purpose we used continuous hashed time-instant private keys (hash chain) in the same way the S/Key system works. Our approach decreases dramatically the number of past time-instant private keys the server stores and only two keys are needed, the last one to construct the previous keys and the first one to recursively verify the authenticity of the next keys.

Keywords: Timed-Release Cryptography, bilinear pairings, S/Key, hash chains, sealed-bid auctions.

1 Introduction

The essence of timed release cryptography (TRC) is to encrypt a message so that it cannot be decrypted by anyone, including the designated recipients, until a specific time-instance. This problem of "sending information into the future" was first mentioned by May [23] in 1993 and then discussed in detail by Rivest et al. [29]. Since its introduction, the solution to the TRC problem has been found useful in a number of real world applications. Some of the best examples are the e-voting which requires delayed opening of votes, the sealed-bid auctions in which the bids must stay sealed so that they cannot be opened before the bidding period and the Internet programming contest where participating teams cannot access the challenge problem before the contest starts. Moreover, TRC can be used for delayed verification of a signed document, such as lottery and check cashing [32] and it can also be applied to online games, especially card games, where players would be able to verify the authenticity of the result when the game ends.

H. Leitold and E. Markatos (Eds.): CMS 2006, LNCS 4237, pp. 130–140, 2006.

1.1 Current TRC Schemes

The existing schemes that solve the TRC problem are divided into two ways – time-lock puzzles [1, 6, 13, 15, 21, 29] and trusted servers [4, 22, 23, 29]. However, none of them are fully satisfactory. Time-lock puzzle approach is based on the required time the receiver needs to perform non-parallelizable computation without stopping. The main advantage of this approach is that no trusted server is needed, but there are also a lot of disadvantages that makes it impractical for real-life scenarios. Some of the drawbacks are that it puts immense computational overhead on the receiver, it depends on the receiver's CPU speed and it does not guarantee that the receiver will retrieve the message immediately after the sender's desired released time have passed. Still, this approach is widely used for specific applications [1, 2, 6, 13, 14, 15, 32].

On the other hand, using trusted servers relieves the receiver from performing non-stop computation and sets the time of the decryption precisely. The cost of this approach is that it requires interaction between the server and the sender or the receiver of the message, or even both. Additionally, some of the existing protocols sacrifice the anonymity of users and sometimes the server is considerably involved in the encryption or decryption process which makes these schemes less scalable. For example there are schemes where the server encrypts messages on request using symmetric encryption and then it publishes the secret key on the designated time. A different scheme was proposed by Di Crescenzo et. al [10], in which non-malleable encryption is used, the server knows nothing about the release time or the identity of the sender, while the receiver engages in a conditional oblivious transfer protocol with the server to get the clear message. Recently, there have been attempts to use bilinear map based IBE schemes for timed release cryptography [3, 4, 9, 22, 28]. Although most of them provide sufficient functionalities, there is still a need of decreasing the amount of data transferred between the users and the server. This is because in a real world application, where a time-server supports thousands or millions of users (including software-agents), there would be 'important' time instances where the majority of receivers will simultaneously try to retrieve the server's private information to read their messages. In this case, a DoS attack may occurred and it is possible that some users will gain advantage through this information.

The main contribution of this paper is to combine the schemes with desirable properties in order to decrease at the minimum the length of the private information that the server reveals and broadcasts at the specific time instance. Our aim is to avoid fairness issues arising from uncontrollable network congestion or delivery delay. Unlike other schemes [3], our approach eliminates the amount of data that the server broadcasts at the designated time and the private information is nothing more than an integer value that is recursively authenticated. Furthermore, we use a continuous hashing procedure to construct the time-instant private keys in the same way the S/Key password authentication system works. Thus, only the current private key is needed to construct the previous keys. The last property is very useful as a user can decrypt messages of previous time-instances by just getting the current private key from the server.

1.2 Properties of a TRC Scheme

To analyze the desirable properties of a TRC scheme, we have to describe some of the applications that require 'future decryption'. As it is referred above, one of these applications is the sealed-bid auction where bidders submit their bids in closed form to the auction board [7, 11, 17, 18, 26, 28]. Once the bidding is closed the bids are opened and the best offer wins the auction. The main problem in the auction is cheating by the auction board or by a competitor. To avoid opening the bids before the desired time, an independent and trusted time-server is needed. To lower the risk that the auction board or a competitor colludes with the time-server, the bidder can use multiple time-servers so that the 'enemy' will need to collude with all the servers to cheat. Some other basic requirements in sealed-bid auctions are that only the auction board will be able to decrypt and verify the bid after the bidding close, the auction board should not be able to disavow bid submission and the bidder should not be able to repudiate his bid.

TRC schemes can also be used to verify the authenticity of the results to the players in an online card game. In this kind of games a user plays against other players or against the gambling company itself. To avoid cheating (from the company) a TRC scheme can be applied together with an independent random generator. In this case a random generator will firstly send the encrypted sequence of cards, so that the players will have the encrypted result before the game starts. When the game ends the time-server will publish the private key for the decryption. Now the players are sure that the company hasn't changed the card sequence during the game. Our approach seems to work very well in this example as a player needs to connect to the time-server only when he stops playing. Then, he gets the private key for the last game and, using continuous hashing, he recursively constructs the private keys for the previous games.

The basic properties of our proposed TRC scheme are:

- The time-server does not interact with either the sender or the receiver.
- The time-instant private key is an integer value (not an Elliptic Curve point [3]) and is also identical for all receivers.
- The public and private key updates published by the time-server inherently authenticate themselves. There is no need of a server signature.
- A Certificate Authority could be used to verify the authenticity of the users.
- The last time-instant private key can be used to construct all the previous time-instant keys.

2 Preliminaries

In the description of our proposed scheme, we will use the following notations and definitions. To better understand the protocol and its security, we review the S/Key system, the bilinear maps and the related mathematical problems we have to face.

2.1 S/Key System

The S/Key one-time password system was proposed by Neil M. Haller [16] in 1995. It is an authentication system which applies a secure hash function multiple times to construct the one-time passwords. The first one-time password is produced by hashing the client's processed password for some specified number of times, say N. The next one-time password is generated by hashing the user's password for only $N - 1$ times. Generally, If f is the hash function, s is the original client's password and $p_{(i)}$ is the one-time password at the i-th login attempt then :

$$p_{(i)} = f^{N-i}(s) \tag{1}$$

This system is secure against eavesdropping attacks as the login – passwords are always different. The eavesdropper cannot produce the next one-time password as the hash function is a one-way function. However, the last property is very useful for the verification of the next password. When the user attempts to login again using the new one-time password, the server checks that the hash product of the new password is equal to the previous password. As there are functions that their hash product is 256 or 512 bits, we can use this procedure in our scheme to construct Elliptic Curve Cryptography private keys of sufficient key-size [20].

2.2 Bilinear Pairings

Suppose $\mathbb{G}_1$ is an additive cyclic group generated by P, whose order is a prime q, and $\mathbb{G}_2$ is a multiplicative cyclic group of the same order. A map $\hat{e}$: $\mathbb{G}_1 \times \mathbb{G}_1 \rightarrow \mathbb{G}_2$ is called a bilinear mapping if it satisfies the following properties:

- Bilinear: $\hat{e}(aP, bQ) = \hat{e}(abP, Q) = \hat{e}(P, abQ) = \hat{e}(P, Q)^{ab}$ for all $P, Q \in \mathbb{G}_1$ and $a, b \in \mathbb{Z}_q^*$
- Non-degenerate: there exist $P, Q \in \mathbb{G}_1$ such that $\hat{e}(P, Q) \neq 1$
- Efficient: there exists an efficient algorithm to compute the bilinear map.

We note that $\mathbb{G}_1$ is the group of points on an elliptic curve and $\mathbb{G}_2$ is a multiplicative subgroup of a finite field. Typically, the Weil, and Tate pairings can be used to construct an admissible bilinear pairing. For a detailed description of pairings and the conditions on elliptic curves one can see [8]. An implementation of the Weil and Tate pairing can be found at [30, 31].

2.3 Mathematical Problems

Definition 1. Discrete Logarithm Problem (DLP)

Given $Q, R \in \mathbb{G}_1$ find an integer $a \in \mathbb{Z}_q^*$ such that $R = aQ$.

Menezes et al. [25] show a reduction from the DLP in $\mathbb{G}_1$ to the DLP in $\mathbb{G}_2$ and they prove that DLP in $\mathbb{G}_1$ is no harder than the DLP in $\mathbb{G}_2$.

Definition 2. Decisional Diffie-Hellman Problem (DDHP)

Given $Q \in \mathbb{G}_1$, aQ, bQ and cQ for some unknowns $a, b, c \in \mathbb{Z}_q^*$ tell whether $c \equiv ab$ (mod q).

While DDHP is hard in $\mathbb{G}_2$, Joux and Nguyen [19] show that DDHP is easy in $\mathbb{G}_1$. Hardness of DDHP in $\mathbb{G}_2$ implies that, $\forall Q \in \mathbb{G}_1$, inverting the isomorphism that takes $P \in \mathbb{G}_1$ and computes $\hat{e}(P, Q)$ is hard [4].

Definition 3. Computational Diffie-Hellman Problem (CDHP)

Given $Q \in \mathbb{G}_1$, aQ, bQ for some unknowns $a, b \in \mathbb{Z}_q^*$, compute abQ.

The advantage of any randomized polynomial-time algorithm $\mathcal{A}$ in solving CDHP in $\mathbb{G}_1$, is defined by the following equation:

$$Adv_{A,G_1}^{CDH} = \text{Prob } [\mathcal{A}(P, aP, bP, abP)=1: a,b \in \mathbb{Z}_q^*] \qquad (2)$$

For every probabilistic algorithm $\mathcal{A}$, Adv_{A,G_1}^{CDH} is negligible.

Definition 4. Bilinear Diffie-Hellman Problem (BDHP)

Given $Q \in \mathbb{G}_1$, aQ, bQ and cQ for some unknowns $a, b, c \in \mathbb{Z}_q^*$, compute $\hat{e}(Q, Q)^{abc}$

If a bilinear pairing exists in the underlying group, the DDHP problem over it can be solved by checking if $\hat{e}(aQ, bQ) = \hat{e}(Q, cQ)$. This lead to the Gap Diffie-Hellman (GDH) assumption according to which, the DDHP on an additive group $\mathbb{G}_1$ can be solved in polynomial time, but there is no polynomial time algorithm to solve the CDHP with non-negligible probability. $\mathbb{G}_1$ is called a GDH group which can be found in supersingular elliptic curves or hyperelliptic curves over finite field. The BDHP over a GDH group is assumed to be difficult and the security of our proposed scheme is based on that assumption.

3 The Proposed TRC Scheme

In this section, we describe the proposed TRC scheme, which is a combination of the scheme proposed by Blake and Chan [3] and the S/Key password authentication system. There is an analysis of its security and an improvement/extension of the way the keys are constructed in order to be better protected against birthday attacks that can be applied in hash functions. [27, 33]. Our encryption scheme is (Gen, TGen, Enc, Dec) for four algorithms such that Gen generates the public and private keys, TGen generates the time-instant keys, Enc encrypts using the receiver's and server's public key, the time-instant public key and the sender's private key and Dec decrypts using the receiver's private key and the time-instant private key.

3.1 General Setup and Key Generation

Let $\mathbb{G}_1$ and $\mathbb{G}_2$ be an additive and multiplicative cyclic group of order q (a prime number) respectively and that $\hat{e}$: $\mathbb{G}_1 \times \mathbb{G}_1 \rightarrow \mathbb{G}_2$ is an admissible bilinear map. The following cryptographic hash functions are chosen: 1) H_1: $\{0, 1\}^* \rightarrow \mathbb{G}_1$, 2) H_2: $\{0, 1\}^*$ $\rightarrow \{0, 1\}^n$ for some n. The notation $H_2^n(x)$ stands for the continuous hashing of x for n times, for example $H_2^3(x) = H_2(H_2(H_2(x)))$. If $n = 0$ then $H_2^0(x) = x$.

There are three entities in the proposed scheme, namely the server (time-server), the sender and the recipient. The server chooses a random private key $s \in \mathbb{Z}_q^*$ and a generator of $\mathbb{G}_1$, say G. The server's public key consists of two elements: G, sG. As for the sender, he chooses a secret private key $a \in Z_q^*$ and he publishes the public key which is: aG, asG. Similarly, the receiver chooses a secret key $b \in Z_q^*$ and he computes the public key: bG, bsG. It is easily understood that this is not an ID-based encryption scheme and a CA type of certification is needed to verify the authenticity of the public keys.

3.2 Time-Instant Key Generation

In our scheme, the construction of the time-instant private keys (the keys needed to decrypt a message at a specified time instance) is based on an S/Key-like procedure. Suppose that the server needs to publish the public keys of a single day and that every key represents a different time instance of that day. To better understand the procedure let us assume that the server needs to publish 24 different public keys and that each one represents a unique hour on that day (eg. 11:00:00 PM Feb 10, 2006 GMT"). For this purpose, the time-server selects a random secret integer value t $\in Z_q^*$. To compute the private key of the first hour of the day T_1 (01:00:00 AM Feb 10, 2006 GMT), the time-server computes the $H_2^{23}(t)$ (this is the private key). The public key for that time instance is: $H_2^{23}(t) \cdot H_1(T_1)$. Similarly, the public key for the next time-instance T_2 (02:00:00 AM Feb 10, 2006 GMT) is $H_2^{22}(t) \cdot H_1(T_2)$ and the same goes for the next time-instances, until the last time instance where the value t is the private key and the point $t \cdot H_1(T_2)$ is the public key. To authenticate a time-instant public key, the server has to publish (together with the public key) the point value $H_2^n(t) \cdot sG$, where $H_2^n(t)$ is the private key for the n-th time instance. To accept a public key, the sender checks if the following equality exists:

$$\hat{e}(H_1(T_n), H_2^{24-n}(t) \cdot sG) = \hat{e}(H_2^{24-n}(t) \cdot H_1(T_n), sG) \tag{3}$$

The trusted time-server publishes the private time-instant key at the specified time.

3.3 Encryption Process

As the public keys consist of two elements we will use the notation Pub_1X to express the first element of the key that belongs to user X and Pub_2X to express the second element e.g for the recipient B : $Pub_1B = bG$ and $Pub_2B = bsG$.

Given a message M, a sender's private key (a), a recipient's public key ($Pub_1B = bG$, $Pub_2B = bsG$), a server's public key ($Pub_1S = G$, $Pub_2S = sG$), a release time $T \in \{0,1\}^*$, and a time-instant public key for the time T: ($Pub_1T = nH_1(T)$, $Pub_2T = nsG$)

1. Verify that $\hat{e}(H_1(T),Pub_2T) = \hat{e}(Pub_1T, Pub_2S) \Rightarrow \hat{e}(H_1(T), nsG) = \hat{e}(nH_1(T), sG) \Rightarrow \hat{e}(H_1(T), G)^{ns} = \hat{e}(H_1(T), G)^{ns}$; if true → accept the time-instant public key.
2. Verify that $\hat{e}(Pub_1B, Pub_2S) = \hat{e}(Pub_1S, Pub_2B) \Rightarrow \hat{e}(bG, sG) = \hat{e}(G, bsG) \Rightarrow \hat{e}(G,G)^{bs} = \hat{e}(G, G)^{bs}$; if true → accept the recipients public key.
3. Choose a random integer $r \in Z_q^*$.

4. Calculate $K = ê(Pub_1T, Pub_1B)^{ar} = ê(nH_1(T), bG)^{ar} = ê(H_1(T), G)^{abnr}$. [1]
5. Send ciphertext $C = < rH_1(T), M \oplus K >$.

3.4 Decryption Process

Given a ciphertext $C = < rH_1(T), M \oplus K >$, a sender's public key (Pub_1A, Pub_2A), a recipients private key (a) and a time-instant private key n,

1. Compute the pairing $K^* = ê(rH_1(T), Pub_1A)^{bn} = ê(rH_1(T), aG)^{bn} = ê(H_1(T), G)^{abnr}$;   if $K^* = K$ then the recipient is sure that the message is not corrupted and he can also verify the sender's identity and validity of the time-instant key n (receiver uses the key n and the sender's public key to compute the pairing $K^* = K$)
2. Recover M by computing $(M \oplus K) \oplus K^* = M$

3.5 A Sketch of Security

To provide a security proof, we work in the same way as Blake and Chan do in [3]. The server's private key s is safe because it is difficult to find s from G, sG (DLP). In the same way, it is difficult to find a user's private key a from G, aG, sG, asG. The argument is as follows: Suppose there exists a polynomial time algorithm $\mathcal{A}(G, aG, sG, asG)$ that finds a. This means that $\mathcal{A}$ can be used to solve the DLP in the following way: Given G, aG, we choose a random integer $b \in Z_q^*$ to compute bG and $baG = abG$; using $\mathcal{A}$, we can find $a = \mathcal{A}(G, aG, bG, baG)$. This problem is as difficult as the DLP.

A message cannot be decrypted since the specified time as the receiver needs to compute $ê(H_1(T),G)^{abnr}$ from sG, aG, asG, b, $rH_1(T)$, $nH_1(T)$ and nsG. As it can been seen, the mapping $ê(H_1(T),G)^{abnr}$ does not contain the server's private key, so the sG, asG and nsG are useless. Suppose that the receiver rewrites G as $wH_1(T)$ for some unknown w, then the problem becomes to find $ê(H_1(T), H_1(T))^{abnrw}$ from $wH_1(T)$, $bH_1(T)$, $rH_1(T)$, $nH_1(T)$ and $awH_1(T)$. This problem is equivalent to the BDHP.

As it can be seen, the easiest way for a receiver to recover a message before the designated time is to solve the Bilinear Diffie-Hellman Problem (BDHP). Hence, as the BDHP problem holds, the recipient cannot gain any information of the encrypted message before its specified release time (excluding the case he colludes with the time-server).

3.6 Extended Private Key Construction

One of the problems that our scheme has to face is that the time-instant private keys are fully dependent. Even though it is very difficult for someone to extract a private key from the public key, we can assume that an attacker finally finds a private key that represents a time instance T_i. Then, by hashing that key, he can produce all the previous private keys $(T_{i-1}, T_{i-2}, ..., T_{i-n})$. This means that he will be able to decrypt all

---

[1] The sender's private key is used in the encryption algorithm in order for the receiver to authenticate the sender's identity during the first step of the decryption process.

the messages sent to him (encrypted with the public keys of these time-instances). To be better protected against this threat, we chose a different procedure for the key construction. Unlike the initial approach, the time-server selects two random secret integer values $t_1, t_2 \in Z_q^*$. The private key for the time-instance T_n is $t_1 \oplus t_2$. The private key for T_{n-1} is $H_2(t_1) \oplus H_2(t_2)$, for T_{n-2} is $H_2^2(t_1) \oplus H_2^2(t_2)$ etc. Using this method of key construction, the knowledge of the private key (of a time instance) does not reveal the private keys that represent previous time-instances.

Although this approach is much safer, it comes with a cost. When the server publishes the private key for a specific time instance he needs also to publish extra information, in order for the users to construct the previous keys. As it is already referred, the main advantage of our protocol is that only the last private key is needed to construct the previous time-instant private keys. For this purpose, if for example the current private key is $H_2(t_1) \oplus H_2(t_2)$, then the server also publishes $H_2(t_2)$. Now a user can compute $H_2(t_1) = (H_2(t_1) \oplus H_2(t_2)) \oplus H_2(t_2)$. Then, he can construct the previous private key $H_2^2(t_1)) \oplus H_2^2(t_2))$.

4   Discussions

To better understand the way our protocol works, we will describe a possible scenario that it can be applied to. Through this scenario, we will discuss a number of desirable properties of the TRC scheme and we will make a comparison with other related schemes.

4.1   Scenario: 'Timed Release Clues'

Suppose a scenario where a user (Bob) wants to send some information (three clues) to a recipient Alice. According to Bob, Alice should not learn all the clues simultaneously, but she can read the clues in a sequential order; each clue at a different time-instance (e.g. after an hour). For this purpose, Bob connects to a time-server that provides public time-instant keys for every single hour of the day and gets the public keys for the time instances $T_1 < T_2 < T_3$.[2] As it was referred in the description of our scheme, the public keys are of the form $(nH_1(T), nsG)$. Bob runs the encryption process of our scheme (the clue$_1$ will be decrypted at T_1, the clue$_2$ at T_2 and the clue$_3$ at T_3) and sends the ciphertexts to Alice. Now, Bob can go offline.

As for the server's side, his first job is to select two random integers t_1, t_2 that will be used for the extended private key construction. For security reasons, the server encrypts and stores these numbers using his public key. The next step is to publish the time-instant public keys. For this purpose, he first creates the private keys and then constructs the public ones.

- for T_3 private: $t_1 \oplus t_2$, public: $(t_1 \oplus t_2)H_1(T_2)$, $(t_1 \oplus t_2)sG$
- for T_2 private: $H_2(t_1) \oplus H_2(t_2)$,   public: $(H_2(t_1) \oplus H_2(t_2))H_1(T_2)$, $(H_2(t_1) \oplus H_2(t_2))sG$

---

[2]   Black and Chan [3] propose another approach where there is no need for a connection to the server to get the public keys.

- for T_1 private: $H_2^2(t_1) \oplus H_2^2(t_2)$, public: $(H_2^2(t_1) \oplus H_2^2(t_2))H_1(T_2)$, $(H_2^2(t_1) \oplus H_2^2(t_2))sG$

When the desirable time comes, the server publishes the private keys together with the needed extra information.

- for T_1 private: $H_2^2(t_1) \oplus H_2^2(t_2)$ extra info: $H_2^2(t_2)$
- for T_2 private: $H_2(t_1) \oplus H_2(t_2)$ extra info: $H_2(t_2)$
- for T_3 private: $t_1 \oplus t_2$ extra info: t_2

The extra information is important to construct the previous time-instant private keys. For example, if the private key for the time instance T_3 has been published, one can also compute the time-instant private key for the time instance T_2 by executing the following operations: get $t_1 = (t_1 \oplus t_2) \oplus t_2$, compute $H(t_1)$ and $H(t_2)$, compute the T_2 private key $\rightarrow H(t_1) \oplus H(t_2)$. Moreover, as the previous private keys can be constructed by the latest published key, the server can delete them from his database. If a user needs to decrypt a message that could have been decrypted on a previous time instance, he just has to get the current key and apply the operation discussed above.

Alice has to wait until the server reveals the private keys. In case where Alice is offline until the time instance T_3, where all the private keys have been revealed, she will get the latest published key (T_3) with its extra information to construct the keys for T_2 and T_1 respectively. This is the main advantage of our protocol. This scheme is very useful for cases where users receive a big number of messages of different time instances. Furthermore, the server does not have to keep lists of previous private keys and at each time the current private key is enough to construct all the previous ones. The communication cost at the decryption process is minimal and the server can support a bigger number of users who simultaneously request the time-instant private keys.

5 Conclusion

In this paper, we described a Timed Release Cryptography scheme that minimizes the connection cost during the receiving process. This scheme can achieve timed release decryption with a precisely specified absolute release time and is scalable enough since the public and private keys are identical for every user. We also provide a solution to missing time-instant private keys by constructing them from the latest published key. Furthermore, all the keys are recursively authenticated without the need of a server's signature.

5.1 Future Work

Currently, we are working on an efficient implementation of the proposed IBE Timed-Release Cryptography scheme. Our aim is to measure the functionality and possible defects of the proposed protocol. We focus on the time-server's resistance to Denial of Service (DoS) attacks and the problems that occur when a big number of users simultaneously request the current private time-instant key. Additionally, we are looking for a stable and safe model that will help us to select an appropriate number for the continuous hashed keys (the length of the hash-chain). As it can be seen, we cannot have a big number of dependent keys, because the scheme will be more vulnerable to birthday attacks. A simple approximation is that if the time-quantum between two release times is very small, we can increase the length of the hash chain; otherwise we decrease it.

Another future research is to use multiple time-servers to lower the risk that a receiver colludes with a time-server. In case where the servers use the same generator G and the clients use the same private key, the problem can be solved as follows: Suppose there are three servers with private keys s_1, s_2 and s_3 respectively. The sender's public keys on each server are (aG, as_1G), (aG, as_2G), (aG, as_3G) and the receiver's public keys are (bG, bs_1G), (bG, bs_2G), (bG, bs_3G). The sender verifies all of the receiver's public keys by checking the equality $\hat{e}(bG, s_xG) = \hat{e}(G, bs_xG)$ for each server s_x. Then, the sender picks a random integer r and calculates the $K_i = \hat{e}(rn_iH_1(T), abG)$ for each $n_iH_1(T)$ (the public time-instant keys for each server). The ciphertext is: $C = < rH_1(T), M \oplus K_1 \oplus \ldots \oplus K_x >$. When the private keys are published, the receiver computes every K_i and then, he is able to decrypt the message.

References

1. M.Bellare and S. Goldwaaser. Encapsulated key escrow. *MIT LCS Tech. Report MIT/LCS/TR-688*, April 1996.
2. M.Bellare and C. Namprempre. Authenticated encryption: Relations among notions and analysis of the generic composition paradigm. In *Proc. Of Asiacrypt '00, Lecture Notes in Computer Science, Vol. 1976*, 2000.
3. I. F. Blake and A. C-F. Chan. Scalable, server-passive, user-anonymous timed release public key encryption from bilinear pairing. *http://eprint.iacr.org/2004/211/*, 2004.
4. D. Boneh and M. Franklin. Identity based encryption from the Weil pairing. In *Advances in Cryptology – Crypto '01, Springer-Verlag LNCS vol. 2139, pages 213-229*, 2001.
5. D. Boneh, B. Lynn, and H. Shacham. Short signatures from the weil pairing. In *Proc. Of Asiacrypt '01*, 2001.
6. D. Boneh and M. Naor. Timed commitments (extended abstract). In *Advances in Cryptology – Crypto 2000, Springer -Verlag LNCS vol. 1880, pages 236-254* 2000.
7. Brandt. Fully private auctions in a constant number of rounds. In *Proceedings of the 7th Annual Conference on Financial Cryptography (FC)*, 2003.
8. J. Cha and J. Cheon. An id-based signature from gap-diffie-hellman groups. In *Public Key Cryptography – PKC 2003*, 2003.
9. L. Chen, K. Harrison, D. Soldera, and N. Smart. Applications of multiple trust authorities in pairing based cryptosystems. In *Proceedings InfraSec 2002, Springer LNCS 2437, pp 260-275*, 2002.
10. G. Di Crescenzo, R. Ostrovsky, and S. Rajagopalan. Conditional oblivious transfer and timed-release encryption. In *Advances in Cryptology –- Eurocrypt '99, Springer-Verlag LNCS vol. 1592, pages 74-89*, 1999.
11. M. K. Franklin and M. K. Reiter. The design and implementation of a secure auction service. In *Proceedings of 1995 IEEE Symposium on Security and Privacy, pp. 2-14, Oakland, California*, 1995.

12. E. Fujisaki and T. Okamoto. Secure integration of asymmetric and symmetric encryption schemes. In *Proceedings CRYPTO 1999, Springer LNCS 1666, pp 537 - 554*, 1999.

13. J. Garay and M. Jakobsson. Timed release of standard digital signatures. In *Financial Crypto '02*, 2002.

14. J. Garay and C. Pomerance. Timed fair exchange of arbitrary signatures. In *Financial Crypto '03*, 2003.

15. J. A. Garay and C. Pomerance. Timed fair exchange of standard signatures. In *Financial Cryptography '02*, 2002.

16. N. Haller. The S/KEY One-Time Password System. *http://www.rfc-archive.org/getrfc.php?rfc=1760*, 2005.

17. J. T. Harkavy and H. Kikuchi. On cheating in sealed-bid auctions. In *EC'03*, 2003.

18. J. T. M. Harkavy and H. Kikuchi. Electronic auctions with private bids. In *3rd USENIX Workshop on Electronic Commerce, Boston, Mass., pp. 61–73*, 1998.

19. A. Joux and K. Nguyen. Separating decision diffie-hellman from diffie-hellman in cryptographic groups. Available from *http://eprint.iacr.org/2001/003/*, 2001.

20. A. K. Lenstra and E. R. Verheul. Selecting Cryptographic Key Sizes. In *Proceedings PKC 2000, Springer- Verlag LNCS 1751, pages 446-465*, 2000.

21. W. Mao. Timed-release cryptography. In *SAC '01, Springer-Verlag LNCS vol. 2259, pages 342-357*, Aug. 2001.

22. K. H. Marco Casassa Mont and M. Sadler. The hp time vault service: Exploiting IBE for timed release of confidential information. In *WWW2003*, 2003.

23. T. May. Timed-release crypto. Manuscript, *http://www.hks.net.cpunks/cpunks-0/1560.html*, Feb. 1993.

24. R.C Mercle. Secure communications over insecure channels. *Communications of ACM, 21(4):294-299*, April 1978.

25. A. Menezes, T. Okamoto, and S. Vanstone. Reducing elliptic curve logarithms to logarithms in a finite field. In *IEEE Transactions on Information Theory IT-39, 5 (1993), 1639–1646*, 1993.

26. M. Naor, B. Pinkas, and R. Sumner. Privacy preserving auctions and mechanism design. In *Proceedings of ACM Conference on Electronic Commerce, pp. 129–139*, 1999.

27. P. van Oorschot, and M. Wiener. A Known Plaintext Attack on Two-Key Triple Encryption. In *Advances in Cryptology – Eurocrypt '90. New York: Springer-Verlag, pp. 366-377*, 1991.

28. I. Osipkov, Y. Kim, and J. H. Cheon. A Scheme for Timed-Release Public Key Based Authenticated Encryption. Available from *http://citeseer.ifi.unizh.ch/709184.html*, 2004.

29. R. L. Rivest, A. Shamir, and D. A. Wagner. Time-lock puzzles and time-released crypto. In *MIT laboratory for Computer Science,MIT/LCS/TR-684*, 1996.

30. Shamus Software Ltd. Miracl: Multiprecision integer and rational arithmetic c/c++ library. Available from *http://indigo.ie/mscott/*.

31. Marcus Stögbauer. Efficient Algorithms for Pairing-Based Cryptosystems. *Diploma Thesis: Darmstadt University of Technology, Dept. of Mathematics*, Jan 2004.

32. P. F. Syverson. Weakly secret bit commitment: Applications to lotteries and fair exchange. In *1998 IEEE Computer Security Foundations Workshop (CSFW11)*, 1998.

33. G. Yuval, How to Swindle Rabin. *Cryptologia 3, pages 187-189*, Jul. 1979.

Compression of Encrypted Visual Data

Michael Gschwandtner, Andreas Uhl, and Peter Wild*

Department of Computer Sciences
Salzburg University, Austria
{mgschwan, uhl, pwild}@cosy.sbg.ac.at

Abstract. Chaotic mixing based encryption schemes for visual data are shown to be robust to lossy compression as long as the security requirements are not too high. This property facilitates the application of these ciphers in scenarios where lossy compression is applied to encrypted material – which is impossible in case traditional ciphers should be employed. If high security is required chaotic mixing loses its robustness to compression, still the lower computational demand may be an argument in favor of chaotic mixing as compared to traditional ciphers when visual data is to be encrypted.

1 Introduction

A significant amount of encryption schemes specifically tailored to visual data types has been proposed in literature during the last years (see [6,10] for extensive overviews). The most prominent reasons not to stick to classical full encryption employing traditional ciphers like AES [3] for such applications are

- to reduce the computational effort (which is usually achieved by trading off security as it is the case in partial or soft encryption schemes),
- to maintain bitstream compliance and associated fuctionalities like scalability (which is usually achieved by expensive parsing operations and marker avoidance strategies), and
- to achieve higher robustness against channel or storage errors.

Compensating errors in transmission of data, especially images, is fundamental to many applications. One example is digital video broadcast or RF transmissions which are also prone to distortions from atmosphere or interfering objects. One famous example for an application scenario requiring security of that type are RF surveillance cameras with their embedded processors, which are used to digitize the signal and encrypt it using state of the art ciphers.

Due to intrinsic properties (e.g. the avalanche effect) of cryptographically strong ciphers (like AES) such techniques are very sensitive to channel errors. Single bits lost or destroyed in encrypted form cause large chunks of data to be lost. Permutations have been suggested to be used in time critical applications since they exhibit significantly lower computational cost as compared to other ciphers, however, this comes at a significantly reduced security level (this is the reason why applying permutations is said

---

* This work has partially been funded by the Austrian Science Fund (FWF), project no. 15170.

H. Leitold and E. Markatos (Eds.): CMS 2006, LNCS 4237, pp. 141–150, 2006.

be a type of "soft encryption"). Hybrid pay-TV technology has extensively used line permutations (e.g. in the Nagravision / Syster systems), many other suggestions have been made to employ permutations in securing DCT-based [11,12,11] or wavelet-based [7,13] data formats. In addition to being very fast, permutations have been identified to be a class of cryptographic techniques exhibiting extreme robustness in case transmission errors occor [9].

The idea of using invertible two-dimensional chaotic maps (CMs) on a square to create symmetric block encryption schemes for visual data is not new and is described in detail in [5] or [2]. Bearing in mind that this type of crypto systems mainly relies on permutations makes them interesting candidates for the use in error-prone environments. Taken this fact together with the very low computational complexity of these schemes, wireless and mobile environments could be potential application fields. In related work we have shown that indeed CMs can cope well with static and random value errors, however, no robustness could be observed with respect to buffer errors since CMs are sensitive to changes in initial conditions.

In this work we focus on an issue different to those discussed so far at first sight, however, this topic is related to the CMs' robustness against value errors: we will investigate the compression of encrypted visual material. Clearly, data encrypted with classical ciphers can not be compressed well: due to the statistical properties of encrypted data no data reduction may be expected using lossless compression schemes, lossy compression schemes can not be employed since the reconstructed material can not be decrypted any more due to compression artifacts. For these reasons, compression is always required to be performed prior to encryption when classical ciphers are used. However, for certain types of application scenarios it may be desirable to perform compression after encryption. CMs are shown to be able to provide this functionality to a certain extent due to their robustness to random value errors. We will experimentally evaluate different CM configurations with respect to the achievable compression rates and quality of the decompressed and decrypted visual data.

A brief introduction to chaotic maps and their respective advantages and disadvantages as compared to classical ciphers will be given in Section 2. Section 3 discusses possible application scenarios requiring compression to be performed after encryption. Experimental results evaluating a JPEG compression with varying quality applied to CM encrypted data are provided in Section 4. Section 5 concludes the paper.

2   Chaotic Map Encryption Schemes

To achieve fast and error-robust encryption of visual data we use CM in form of a permutation based symmetric cipher. This approach was originally introduced by the work of F. Pichler and J. Scharinger [8] and has been extended by J. Fridrich [5]. CM encryption relies on the use of discrete versions of chaotic maps. The good diffusion properties of chaotic maps, such as the *Bakermap* or *Catmap* soon attracted cryptographer. Turning a chaotic map into a symmetric block cipher requires three steps, as [5] points out.

1. **Generalization:** Once the chaotic map is chosen, it is desirable to vary its behavior through parameters. These are part of the *key* of the cipher.

2. **Discretization:** Since chaotic maps usually are not discrete, a way must be found to apply the map onto a finite square lattice of points that represent pixels in an invertible manner.
3. **Extension to 3D:** As the resulting map after step two is a parameterized permutation, an additional mechanism is added to achieve substitution ciphers. This is usually done by introducing a position-dependent gray level alteration.

In most cases a final **diffusion step** is performed, often achieved by combining the data line or column wise with the output of a random number generator.

The most famous example of a chaotic map is the standard *Bakermap*:

$$B : [0, 1]^2 \rightarrow [0, 1]^2.$$

$$B(x, y) = \begin{cases} (2x, \frac{y}{2}) & \text{if } 0 \leq x < \frac{1}{2}; \\ (2x - 1, \frac{y+1}{2}) & \text{if } \frac{1}{2} \leq x \leq 1. \end{cases}$$

This geometrically corresponds to a division of the unit square into two rectangles $[0, \frac{1}{2}[\times [0, 1]$ and $[\frac{1}{2}, 1] \times [0, 1]$ that are stretched horizontally and contracted vertically. Such a scheme may easily be generalized using k vertical rectangles $[F_{i-1} F_i[\times [0, 1[$ each having an individual width p_i such that $F_i = \sum_{j=1}^{i} p_j$, $F_0 = 0$, $F_k = 1$. The corresponding vertical rectangle sizes p_i, as well as the number of iterations, are introduced as parameters. Another choice of a chaotic map is the *Arnold Catmap*:

$$C : [0, 1]^2 \rightarrow [0, 1]^2.$$

$$C(x, y) = \begin{pmatrix} 1 & 1 \\ 1 & 2 \end{pmatrix} \begin{pmatrix} x \\ y \end{pmatrix} \mod 1$$

where $x \mod 1$ denotes the fractional part of a real number x by subtracting or adding an appropriate integer. This chaotic map can be generalized using a matrix A introducing two integers a, b such that $det(A) = 1$ as follows:

$$C_{gen}(x, y) = A \begin{pmatrix} x \\ y \end{pmatrix} \mod 1, \ A = \begin{pmatrix} 1 & a \\ b & ab + 1 \end{pmatrix}.$$

Now each generalized chaotic map needs to be modified to turn into a bijective map on a square lattice of pixels. Let $\mathcal{N} := \{0, \dots, N - 1\}$, the modification is to transform domain and codomain to $\mathcal{N}^2$. Discretized versions should avoid floating point arithmetics in order to prevent an accumulation of errors. At the same time they need to preserve sensitivity and mixing properties of their continuous counterparts. This challenge is quite ambitious and many questions arise, whether discrete chaotic maps really inherit all important aspects of chaos by their continuous versions. An important property of a discrete version F of a chaotic map f is:

$$\lim_{N \to \infty} \max_{0 \leq i, j < N} |f(i/N, j/N) - F(i, j)| = 0.$$

To give an example, discretizing a chaotic *Catmap* is fairy simple and introduced in [2]. Instead of using the fractional part of a real number, the integer modulo arithmetic is adopted:

$$C_{disc} : \mathcal{N}^2 \rightarrow \mathcal{N}^2.$$

$$C_{disc}(x, y) = A \begin{pmatrix} x \\ y \end{pmatrix} mod\ N, \ A = \begin{pmatrix} 1 & a \\ b & ab + 1 \end{pmatrix}$$

Finally, an *extension to 3D* is inserted, that may be applied to any two-dimensional chaotic map. As all chaotic maps preserve the image histogram (and with it all corresponding statistical moments) a procedure to result in a uniform histogram after encryption is desired. The extension of a two dimensional discrete chaotic map $F : \mathcal{N}^2 \to \mathcal{N}^2$ to three dimensions consists of a position dependent gray-level shift (assuming L gray-levels $\mathcal{L} := \{0, \ldots, L - 1\}$) at each level of iteration:

$$F_{3D} : \mathcal{N}^2 \times \mathcal{L} \to \mathcal{N}^2 \times \mathcal{L}$$

$$F_{3D}(i, j, g_{ij}) = \begin{pmatrix} i' \\ j' \\ h(i, j, g_{ij}) \end{pmatrix}, \ \begin{pmatrix} i' \\ j' \end{pmatrix} = F(i, j).$$

The map h modifies the gray-level of a pixel and is a function of the initial position and color of the pixel, that is $h(i, j, g_{ij}) = g_{ij} + \overline{h}(i, j)\ mod\ L$. There are various possible choices of $\overline{h}$, we use $\overline{h}(i, j) = i \cdot j$.

Chaotic maps after step two or three are bijections of a square lattice of pixels. An additional spreading of local information over the whole image is desirable. Otherwise the cipher is vulnerable to *Known Plaintext Attacks*, since each pixel in the encrypted image corresponds to exactly one pixel in the original. The diffusion step is often realized as a line-wise process, e.g.

$$v(i, j)^* = v(i, j) + G(v(i, j - 1)^*)\ mod\ L$$

where $v(i, j)$ is the not-yet modified pixel at position (i, j), $v(i, j)^*$ is the modified pixel at that position, and G is a random look-up table.

Concerning robustness against transmission or storage errors, it is of course better to avoid diffusion steps. If local information is spread during encryption, i.e. in diffusion steps, a single pixel error in the encrypted image causes several pixel errors in the original image. For this reason we investigate both settings, with and without diffusion.

3   Application Scenarios: Compression and Encryption

As already outlined in the introduction, classically encrypted images normally can not be compressed very well (actually these data should not be compressible at all), because of the typical properties encryption algorithms have. In particular it is not possible to employ lossy compression schemes since in this case potentially each byte of the encrypted image is changed (and most bytes in fact are), which leads to the fact that the decrypted image is entirely destroyed resulting in a noise-type pattern. Therefore, in all applications involving compression and encryption, compression is performed prior to encryption.

On the other hand, application scenarios exist where a compression of encrypted material is desirable. In such a scenario classical block or stream ciphers cannot be employed. For example, dealing with video surveillance systems, often concerns about protecting the privacy of the recorded persons arise. People are afraid what happens with recorded data allowing to track a persons daily itineraries. A compromise to minimize impact on personal privacy would be to continuously record and store the data but only view it, if some criminal offence has taken place.

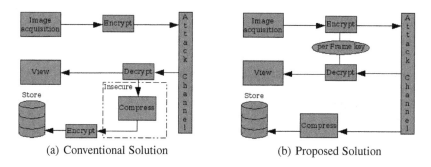

(a) Conventional Solution   (b) Proposed Solution

Fig. 1. Privacy Solutions for Surveillance-Systems

To assure, that data can not be reviewed unauthorized, it is transmitted and stored in encrypted form and only few people have the authorization (i.e. the key material) to decrypt it.

The problem, as depicted in Figure 1.a , is the amount of memory needed to store the encrypted frames (due to hardware restrictions of the involved cameras, the data is transmitted in uncompressed form in many cases). For this reason, frames should be stored in a compressed form only. When using classical ciphers the only way to do this would be the decryption, compression and re-encryption of frames. This would allow the administrator of the storage device to view and extract the video signal which obviously threatens privacy. There are two practical solutions to this problem:

1. Before the image is encrypted and transmitted, it is compressed. Beside the above-mentioned computational demands for the camera system, this has further disadvantages, as transmission errors in compressed images have usually an even bigger impact. This is prohibitive in environments where the radio signal is easily distorted.

2. The encrypted frames are compressed directly. In this manner, the key material does not have to be revealed when storing the visual data thereby maintaining the privacy of the recorded persons. Figure 1.b shows such a system. Clearly, in this scenario classical encryption cannot be applied. In the following we will investigate whether CM can be applied and which results in terms of quality and compression are to be expected.

A second example where compression of encrypted visual data is desirable is data transmission over heterogenous networks, for example a transition from wired to wireless networks with corresponding decreasing bandwidth. Consider the transmission of uncompressed encrypted visual data in such an environment – when changing from the wired network part to the wireless one, the data rate of the visual material has to be reduced to cope with the lower bandwidth available. Employing a classical encryption scheme, the data has to be decrypted, compressed, and re-encrypted, similar to the surveillance scenario described before. In the network scenario these operations put significant computation load onto the network node in charge for the rate adaptation **and** the key material needs to be provided to that network node, which is demanding in terms of key management. A solution where the encrypted material may be compressed directly is much more efficient of course.

4   Compressing CM Encrypted Images

4.1   Experimental Setup

We present results of four different flavours of the chaotic *CatMap* algorithm (the results concerning the *Bakermap* are very similar, therefore we only provide results for two variants) (see Table 1). The diffusion step has been excluded from all chaotic maps, except *CatDiff*. All algorithms are applied to a 256×256 version of the *Lena* test image with 256 gray levels using two sets of representative encryption keys.

Table 1. Tested image encryption algorithms

Name	Description
2DCatMap5	Catmap with five iterations.
2DCatMap10	Catmap with ten iterations.
2DCatDiff5	Catmap with diffusion step and five iterations.
3DCatMap5	Catmap with 3D extension and five iterations.
2DBMap5	Bakermap with five iterations.
2DBMap17	Bakermap with seventeen iterations.

After encryption, JPEG compression is applied to the encrypted image data. To assess the behaviour of the described processing pipeline, the image is finally decompressed, decrypted and the result is compared to the original image and the achieved compression ratio is recorded. Note that it is difficult to find reliable tools to measure quality of distorted images, especially in a low-quality scenario. Several metrics exist, such as the Signal to Noise Ratio (SNR), Peak SNR (PSNR) or Mean Square Error (MSE), which are frequently used in quantifying distortions (see [4,1]). However, reliable assessment of low quality images should be made by human observers in a subjective rating as this can not be accomplished in a sensile way using the metrics above. It is clear that these measurements are time consuming, as they can not be automated. In order to complement the visual examples, we also report the reference PSNR value.

4.2   Experimental Results

Figs. 2 – 5 show images where the encrypted data got lossy (JPEG) compressed, decompressed and finally decrypted again. In these figures, we provide the quality factor q of the JPEG compression, the data size of the compressed image in percent % of the original image size, and the PSNR of the decompressed and decrypted image given in dB.

In general, we observe quite unusual behavior of the CM encryption technique. The interesting fact is that despite the lossy compression a CM encrypted image can be decrypted quite well (depending on the compression rate of course). As already mentioned, this is never the case if classical encryption is applied.

Fig. 2 compares the application of the standard 2D Catmap without and with additional extensions to increase security (i.e. 3D or diffusion extensions are employed additionally). At a fixed compression rate (slightly lower than 3) we obtain a somewhat noisy but clearly recognizable image in case of no further extensions are used (Fig. 2.a). Applying the 3D extension to the standard Catmap (Fig. 2.b), we observe significant degradation of the decrypted image as compared to the standard Catmap with identical number of iterations. However, the image content is still recognizable which is no longer true in case the diffusion extension is used – see Fig.2.c. It is worthwhile noticing that we obtain the same result – noise – no matter which compression rate or image quality is used in case the diffusion step is performed. Actually this result is identical to a result if AES had been used instead of *Catdiff*.

(a) q=55: 36%, 23.4dB   (b) q=45: 37%, 15.9dB   (c) q=45: 37%, 9.2dB

Fig. 2. Catmap with 5 iterations (without extensions and using 3D and diffusion extensions, respectively), keyset 1

The effect when compression ratio is steadily increased is shown in Fig. 3. Lower data rates in compression increase the amount of noise in the decrypted images, however, still with a compression ratio of 5 (20%) the image is clearly recognizable and the quality would be sufficient for a handheld phone or PDA display for example (Fig. 3.b). Of course, higher compression ratios lead to even more severe degradations which are hardly acceptable for any application (e.g. compression ratio 8 in Fig. 3.c).

Increasing the number of iterations to more than 5 does not affect the results of the Catmap for a sensible keyset (as used for example in Fig. 3). This is not true for the Bakermap as shown in Fig. 4. When using 5 iterations, the compression result is significantly better as compared to the Catmap case with the same data rate (compare Fig. 4.a to Fig. 2.a). The reason is displayed in Fig. 4.b – using the Bakermap with 5 iterations, we still recognize structures in the encrypted data which means that mixing has not yet fulfilled its aim to a sufficient degree. On the one hand, this is good for compression

(a) q=35: 27%, 19.9dB   (b) q=25: 20%, 17.9dB   (c) q=15: 12%, 16.4dB

Fig. 3. Catmap with 5 iterations using different compression ratios, keyset 2

(a) q=70: 37%, 28.0dB   (b) q=70: encrypted   (c) q=60: 36%, 24.9dB   (d) q=60: encrypted

Fig. 4. Bakermap with varying number of iterations (5 and 17 iterations), keyset 2

since errors are not propagated to a large extent, on the other hand this threatens security since the structures visible in the encrypted data can be used to compute key data used in the encryption process.

Increasing the number of iterations (e.g. to 17 as shown in Figs. 4.c and 4.d) significantly reduces the amount of visible structures. As it is expected, the compression results are similar now to the Capmap case using 5 iterations. Using 20 iterations and more, no structures are visible any more and the compression results are identical to the Catmap case.

In Fig. 5 we give examples of the effects in case pathological key material is used for encryption. When using keyset 1 for encryption with the Bakermap (Figs. 5.a and 5.b), the structures visible in the encrypted material are even clearer and in perfect correspondence also the compression result is superior to that of keyset 2 (Fig. 4). With these setting, an even higher number of iterations is required to achieve reasonable security (which again destroys the advantage with respect to compression). Also for the Catmap, weak keys exist. In Fig. 5.d the encrypted data is shown in case 10 iterations are performed using keyset 1. In this case, even image content is revealed and the key parameters are reconstructed easily with a ciphertext only attack. Correspondingly, also the compression results are much better as compared to the case when 5 iterations are applied (see Fig. 2.a). These parameters (weak keys) and effects (reduced security) have been described in the literature on CM and have to be avoided for any application of course.

(a) q=75: 36%, 30.9dB   (b) q=75: encrypted   (c) q=70: 36%, 27.3dB   (d) q=70: encrypted

Fig. 5. Bakermap and Catmap with pathological keys (5 and 10 iterations)

In general, we observe a significant tradeoff between security and visual quality of compressed data when comparing the different settings as investigated. Increasing the number of iterations up to a certain level increases security but decreases compression performance (this is especially true for the Bakermap which requires a higher number of iterations in general to achieve reasonable security). However, of course the computational effort increases as well.

We face an even more significant tradeoff when increasing security further – the 3D extensions already strongly decrease image quality whereas diffusion entirely destroys the capability of compressing encrypted visual data. When the security level approaches the security of cryptographically strong ciphers like AES, also CMs do not offer robustness against lossy compression any longer.

5   Conclusion

Chaotic mixing based encryption techniques are shown to tolerate a medium amount of lossy compression which is an exceptional property not found in other ciphers. Applying the Catmap with 5 iterations or the Bakermap with 20 iterations provides reasonable security and decrypted images show acceptable image quality even after significant JPEG compression. However, if techniques enhancing CMs security like the 3D extension technique or diffusion are used, the robustness against compression is reduced or entirely lost.

As long as a lower security level is desired or acceptable (i.e. 3D extension or diffusion is omitted), CM may be employed in application scenarios where lossy compression is applied to the encrypted data. This type of application scenarios cannot be operated with traditional ciphers. If high security is required (and the lower computational demand of CM is not an issue) it is better to stick to classical block ciphers in any environment since CM loses its robustness to compression anyhow.

References

1. I. Avcibas, B. Sankur, and K. Sayood. Statistical evaluation of image quality measures. *Journal of Electronic Imaging*, 11(2):206–223, April 2002.
2. G. Chen, Y. Mao, and C.K. Chui. A symmetric image encryption scheme based on 3D chaotic cat maps. *Chaos, Solitons and Fractals*, 21:749–761, 2004.

3. J. Daemen and V. Rijmen. *The Design of Rijndael: AES — The Advanced Encryption Standard*. Springer Verlag, 2002.
4. A. Eskicioglu. Quality measurement for monochrome compressed images in the past 25 years. In *Proceedings of the International Conference on Acoustics, Speech and Signal Processing*, pages 1907–1910, 2000.
5. J. Fridrich. Symmetric ciphers based om two-dimensional chaotic maps. *Int. J. of Bifurcation and Chaos*, 8(6):1259–1284, 1998.
6. B. Furht and D. Kirovski, editors. *Multimedia Security Handbook*. CRC Press, Boca Raton, Florida, 2005.
7. R. Norcen and A. Uhl. Encryption of wavelet-coded imagery using random permutations. In *Proceedings of the IEEE International Conference on Image Processing (ICIP'04)*, Singapore, October 2004. IEEE Signal Processing Society.
8. J. Scharinger. Fast encryption of image data using chaotic Kolmogorov flows. *Journal of Electorinic Imaging*, 7(2):318–325, 1998.
9. Ali Saman Tosun and Wu chi Feng. On error preserving encryption algorithms for wireless video transmission. In *Proceedings of the ninth ACM Multimedia Conference 2001*, pages 302–307, Ottawa, Canada, October 2001.
10. A. Uhl and A. Pommer. *Image and Video Encryption. From Digital Rights Management to Secured Personal Communication*, volume 15 of *Advances in Information Security*. Springer-Verlag, 2005.
11. Jiangtao Wen, Mike Severa, Wenjun Zeng, Max Luttrell, and Weiyin Jin. A format-compliant configurable encryption framework for access control of video. *IEEE Transactions on Circuits and Systems for Video Technology*, 12(6):545–557, June 2002.
12. W. Zeng, J. Wen, and M. Severa. Fast self-synchronous content scrambling by spatially shuffling codewords of compressed bitstreams. In *Proceedings of the IEEE International Conference on Image Processing (ICIP'02)*, September 2002.
13. Wenjun Zeng and Shawmin Lei. Efficient frequency domain selective scrambling of digital video. *IEEE Transactions on Multimedia*, 5(1):118–129, March 2003.

Selective Encryption for Hierarchical MPEG

Heinz Hofbauer, Thomas Stütz*, and Andreas Uhl

University of Salzburg,
Department of Computer Sciences
{hhofbaue, tstuetz, uhl}@cosy.sbg.ac.at

Abstract. Selective encryption of visual data and especially MPEG has attracted a considerable number of researchers in recent years. Scalable visual formats are offering additional functionality, which is of great benefit for streaming and networking applications. The MPEG-2 and MPEG-4 standards provide a scalability profile in which a resolution scalable mode is specified. In this paper we evaluate a selective encryption approach on the basis of our hierarchical MPEG video codec.

1 Introduction

Encryption schemes for visual data need to be specifically designed to protect the content while preserving properties of its representation in the encrypted domain. Furthermore the real-time encryption of a video stream with state-of-the-art ciphers still requires heavy computation, especially when considering target hardware platforms like set-top boxes for digital TV broadcasts. Numerous attempts have been made to secure MPEG streams, among them [1–4]. Selective encryption has been accomplished in various ways, encryption of I-frames, motion vector data, coefficient permutation, ... Several approaches do not strive for maximum security, but trade off security for computational complexity. For a detailed discussion please refer to [5]. Also the JPEG standard and its scalable modes of operation were target of research concerning selective encryption [6, 7]. The rising importance of scalability preserving encryption of scalable video streams has been discussed in [8–10]. Selective encryption of scalable video streams can greatly reduce the complexity of video distribution in different qualities/resolutions. If the scalability is preserved in the encrypted domain, no decryption key, no decryption, no transcoding and no reencryption is necessary for accessing the lower resolution versions of a stream. The high quality layers can simply be dropped e.g., by a simple network set-top box. The

---

* The support of the Austrian Grid project is gratefully acknowledged.

** The work described in this paper has been supported in part by the European Commission through the IST Programme under Contract IST-2002-507932 ECRYPT. The information in this document reflects only the author's views, is provided as is and no guarantee or warranty is given that the information is fit for any particular purpose. The user thereof uses the information at its sole risk and liability.

*** This work has been partially supported by the Austrian Science Fund FWF, project number 15170.

H. Leitold and E. Markatos (Eds.): CMS 2006, LNCS 4237, pp. 151–160, 2006.

computational effort is to parse the code stream for marker sequences to identify the relevant parts, which is negligibly small.

The paper presents a selective encryption approach for a hierarchical MPEG coder and evaluates its performance. An advantage of using scalable video codecs is that the overhead to identify the relevant parts for encryption can be kept very low and that rate adaption can easily be conducted in the encrypted domain. In section 2 we give an overview of the hierarchical MPEG coder. The selective encryption approach is presented in section 3. Empirical results of this selective encryption approach are discussed in section 4 and section 5 concludes the paper.

2   Hierarchical MPEG

HMPEG (Hierarchical MPEG) is closely related to the MPEG resolution scalable mode as defined for MPEG-2 and MPEG-4 Part 2. Since there is no freely available implementation capable of compressing in this mode (the scalable modes are altogether poorly supported) we had to implement it from the scratch. The compression performance of the HJPEG coder is very close to baseline JPEG [6]. The MPEG-2 code stream syntax is rather complex and of no special interest for our investigations, thus our implementation does not produce a standard-compliant MPEG-2 stream but has all its properties. Hence all our results are also applicable to resolution scalable MPEG-2. The MPEG standards basically apply motion compensation to exploit temporal redundancy and compress the resulting frames in a way very similar to JPEG. Our implementation employs hierarchical JPEG as defined in [11] for frame compression.

2.1   Hierarchical JPEG

HJPEG (Hierarchical JPEG) is a resolution scalable compression method, which is part of the JPEG standard [11]. A number of layers is chosen and for each layer the image is downsampled by a factor of two in each dimension with up-sampling and downsampling filters as proposed in [11]. The reconstruction of a certain resolution is used as a prediction for the next higher resolution. The re-sulting series of difference frames and the lowest resolution subsample are JPEG encoded. The HJPEG compression is conducted with a custom implementation based on the Independent JPEG Group's library libjpeg.

2.2   MPEG Video Compression

For motion compensation a frame is split up into macroblocks (16x16 pixel in MPEG-2 and our implementation). For each of these macroblocks the best matching block in another frame is located (in our implementation with full pixel accuracy) and the difference calculated. Additionally there is the possibility that no good enough macroblock can be found, resulting in a so called I-macroblock which contains original image information. Motion compensation is accompa-nied by a still image compression system that applies a DCT (discrete cosine transform) and a Huffman based entropy coder. MPEG-2 uses three different frame types:

- I-Frame (Intra Frame): contains solely original image information.
- P-Frame (Predicted Frame): contains the difference between the previous and the actual frame.
- B-Frame (Bidirectionally Predicted Frame): uses the previous and next I- or P-frame for the computation of the prediction.

The repeated structure of these frames is called group of pictures (GOP) and has to start with an I-frame, e.g., I B B P B B P.

Putting all together our HMPEG implementation performs motion compensation on the basis of 16x16 pixel macroblocks with a one pixel accuracy. HJPEG is used to compress the frames in a resolution scalable fashion delivering a resolution scalable video stream.

3   Selective Encryption

One goal of selective encryption is the preservation of the scalability in the encrypted domain. Thus no key, no decryption, no transcoding and reencryption is necessary to access lower resolution versions of the video stream, as it would be the case for regular encryption and non-scalable video data. Another possible goal is the reduction of the encryption complexity by reducing the amount of data to be encrypted. Application scenarios for selective encryption can be divided into two groups.

Confidential encryption has the same goal as regular encryption, the secure scrambling of all image information.

Transparent encryption is here used as an umbrella term for all application scenarios where confidential encryption is not demanded. These application scenarios may impose two requirements:

- Security requirement: a certain portion of the visual information has to be securely removed.
- Quality requirement: a certain image quality may have to be preserved.

A content provider might want to reveal a low quality version in order to attract costumers and lure them into buying the high quality version. Another case is e.g., the transmission of a soccer game. The information that a soccer game is broadcasted is not subject to secrecy, but nevertheless the broadcast should only be consumable by paying customers, that can decipher the encrypted broadcast. Here the security requirement is to achieve sufficiently bad quality.

3.1   Approach

The targets of our encryption approach are the compressed coefficient data of the frames and the motion vectors. Therefore we have to identify these parts. In our implementation we had to deal with the JPEG code stream syntax but essentially the same is possible with MPEG-2 code stream syntax. For the JPEG code stream syntax we have to parse for markers indicating a scan, the JPEG code stream unit containing only compressed coefficient data. After encryption with a state-of-the-art cryptographic cipher (AES [12]) we apply byte stuffing to avoid maker sequences and to preserve the scalability. The code stream syntax is at least partially preserved except for valid Huffman codes. In the case of MPEG the parsing of the relevant parts is different. In MPEG-2 these marker sequences are called start codes and consist of a 3 byte prefix (0x00 00 01) followed by a 1 byte identifier. The relevant parts (compressed coefficient data and motion vectors) have to be identified (slices in MPEG) and the generation of start codes prevented.

3.2 Attacks

The proposed encryption approach results in a pseudo code-stream-compliant file which can be at least partly recovered by a code-stream-compliant decoder. However, this reconstruction will have a significantly lower quality than a possible attacker can achieve. The reason is that our method introduces heavy random

P-frame with encrypted MVs   P-frame after an encrypted I-frame

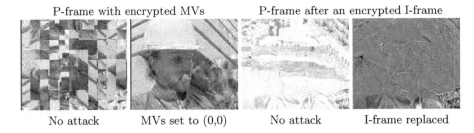

No attack   MVs set to (0,0)   No attack   I-frame replaced

Fig. 1. Replacement attacks versus the reconstruction of the partially encrypted Foreman video

distortion which also greatly reduces the image quality of unencrypted parts of the video stream. An attacker can replace encrypted parts with data that does not introduce random noise. This method is called replacement attack. In our case encrypted I-frames are replaced by a uniform gray image. The encrypted difference frames are replaced by zero valued frames.

If the motion vectors are encrypted the replacement attack is to either set them to $(0,0)$ or to decode each difference frame to obtain the high frequency changes between frames. Figure 1 illustrates the replacement attack for various coding and encryption settings.

4   Results

HMPEG with two HJPEG layers was applied and all results are obtained by conducting a replacement attack. Two layers were used because the resolution of the Foreman test sequence has been too low (176x144) to justify more layers. In this paper only the PSNR plots are given, but also the LSS/ESS values [13] were evaluated. The LSS/ESS plots basically show the same behavior as the PSNR plots and therefore only the more familiar PSNR plots are given.

4.1   Confidential Encryption

All HJPEG layers contain at least high frequency data that reveals information about the image content. Figure 2 reveals that this also holds for predicted frames (only a minor adjustment of the color levels was performed). Therefore - and on the basis of many other evaluations - one can state that nearly all of the HMPEG stream has to be encrypted to achieve perfect secrecy. Even the motion vectors alone reveal enough image information to roughly guess the content.

2nd layer of 9th P-frame   Original picture

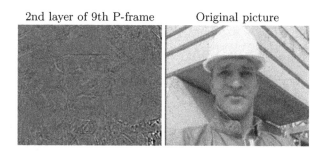

Fig. 2. Image information in a HJPEG layer of a predicted frame

4.2   Transparent Encryption

The requirements of transparent encryption may vary from application to application. The following results will hopefully give the reader some insight into what can be achieved with selective encryption of HMPEG. An application that needs to preserve a certain image quality only keeps the resolutions unencrypted that satisfy the quality requirement. In the following we tried to analyze how the encryption effort can be minimized while severely distorting the video quality. We evaluated the image quality for various encryption and coding settings, including different GOP structures.

Figure 3 shows the PSNR plot of the first 125 frames (GOP: one I-frame and the rest P-frames) of the Foreman sequence with various parameter and encryption settings. Only the I-frame has been encrypted (both layers). Encrypted motion vectors are indicated by the label `mv` and the usage of I-macroblocks is indicated by the label `imb`. To take into account that the addition of difference frames without a reference I-frame may also introduce additional distortion, we also plotted the PSNR of the layer 1. This layer 1 PSNR plot (`layer1direct`) extracts the highest resolution HJPEG layer of each frame and uses it directly. Since the highest resolution HJPEG layer produces the biggest amount of HMPEG data, encrypting it leads to the encryption of most of the HMPEG stream. In this sense the `layer1direct` PSNR graph shows a lower bound for the video quality while only encrypting a rather small portion of the overall HMPEG stream data.

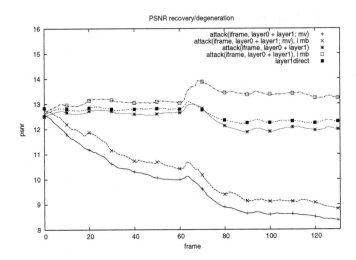

Fig. 3. PSNR of the HMPEG compressed Foreman sequence for a GOP with only P-frames, I-frame encrypted

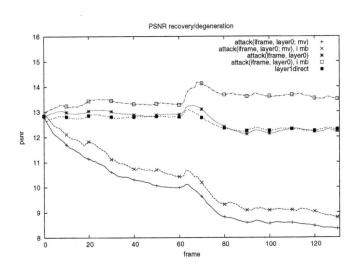

Fig. 4. PSNR of the HMPEG compressed Foreman sequence for a GOP with only P-frames, base layer of I-frame encrypted

The original image, an estimation of the image via replacement attack of the I-frame (no I-macroblocks, unencrypted motion vectors) and the direct use of layer 1 is illustrated in Fig. 5 for the 100th P-frame of the Foreman sequence. Furthermore, should the PSNR of a sequence generated by a replacement attack be lower than the layer1direct sequence the partial encryption is sufficient not only to remove all information of the I-frame but also to prevent any noticeable

Original frame   Replacement attack   Layer 1 direct

Fig. 5. Reconstructions based on two attacks compared to the original image for the Foreman sequence for the 100th P-frame (GOP with only I-frames)

regenerative effect induced by the P-frames. Figure 3 states that the encryption of the I-frame is sufficient for destroying the visual quality of the whole sequence. Furthermore, when the motion vectors are encrypted the quality decreases even more, due to the addition of the difference frames to wrong spatial locations of the image. Figure 6 illustrates the restricted plausibility of the PSNR. The 50th P-frame of the `layer1direct` sequence, the sequence with the I-frame encrypted and the sequence with I-frame and motion vectors encrypted. Encrypting and attacking the motion vectors really degrades the quality as the PSNR plot would suggest.

Nevertheless if the motion vectors are left unencrypted, the quality of the image is better than its PSNR value indicates. While the PSNR of the attacked frame is lower than the PSNR of the layer 1 of the frame, certain details are better visible in the attacked frame and there is even some regeneration regarding the texture.

Layer 1 direct   Encrypted I-frame   Encrypted I-frame & MVs

Fig. 6. The 50th P-frame of the Foreman sequence for different encryptions and attacks, no I-macroblocks are used (GOP with only P-frames)

Regardless of these problems the PSNR plot of the Foreman sequence is a typical one. Unsurprisingly the use of I-macroblocks raises the quality of the frames, though without changing the basic nature of the results. The encryption of the motion vectors helps to prevent the sequence from regenerating and `layer1direct` yields to a better result than the replacement attack. The sequence with only the I-frame encrypted is better than the layer 1 used directly, although the plot suggests otherwise. Interestingly the encryption of only the base layer of an I-frame yields to very similar results. (see Fig. 4). This means only a fraction of the I-frames data needs to be encrypted to degenerate the quality in a way quite similar to encrypting the whole I-frame. The amount of movement in the sequence influences the gain of image quality when using I-macroblocks. In sequences where there is little movement, I-macroblocks will seldom come to bear since the difference is mostly caught by the motion vectors. Such a case is illustrated in Fig. 7 for the Akiyo sequence. If the GOP is

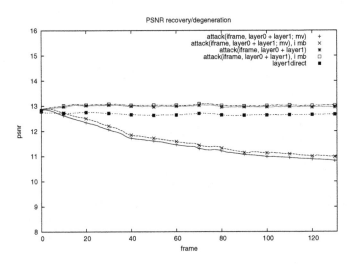

Fig. 7. PSNR of the HMPEG compressed Akiyo sequence with a GOP of only P-frames, I-frame encrypted

changed to IBBPBBP... the overall behavior stays the same but the quality becomes slightly better. This is due to the higher distance between P-frames, which results in difference frames containing more information. For the same reason the influence of I-macroblocks with this GOP is higher, since the greater difference between the frames leads to a higher number of I-macroblocks. Figure 8 shows a PSNR plot for the Foreman sequence using a GOP of IBBPBBP.... . Tough PSNR would suggest a noticeable increase in the visual quality this higher PSNR only reflects the higher number of I-macroblocks. For an IBBPBBP... GOP using I-macroblocks the 50th image of an attacked sequence is depicted in Fig. 9. The visual quality is still poor and has not changed much, except for the visible I-macroblocks, regarding a GOP with one I-frame and the rest P-frames (c.f. Fig. 6).

Again very similar results are obtained by encrypting only the base layer of the I-frame.

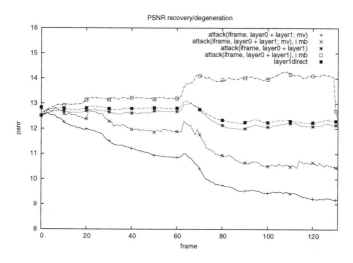

Fig. 8. PSNR of the HMPEG compressed Foreman sequence with a GOP of IBBPBBP..., I-frame encrypted

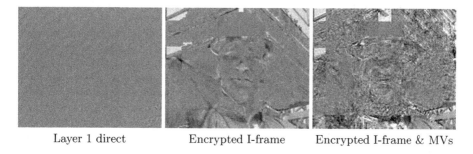

| Layer 1 direct | Encrypted I-frame | Encrypted I-frame & MVs |

Fig. 9. The 50th P-frame of the Foreman sequence for different encryptions and attacks, I-macroblocks are used (GOP with P- and B-frames)

5   Conclusion

In this paper we presented a selective encryption approach for HMPEG that is capable of preserving the scalability in the encrypted domain. Furthermore we evaluated its suitability for confidential and transparent encryption. For confidential encryption most of the video stream has to be encrypted including motion vectors. Nevertheless the scalability is fully preserved even on a JPEG scan basis for our implementation or on a slice basis for MPEG. Furthermore we could show that it is possible to severely degrade the video quality by only encrypting very little of the overall video stream data (about 0.3% for 125 P- or B-frames in a GOP). It is sufficient to encrypt the base layer of all I-frames in order to severely distort and reduce the video quality.

References

1. Kunkelmann, T.: Applying encryption to video communication. In: Proceedings of the Multimedia and Security Workshop at ACM Multimedia '98, Bristol, England (1998) 41–47
2. Dittmann, J., Steinmetz, R.: A technical approach to the transparent encryption of MPEG-2 video. In Katsikas, S.K., ed.: Communications and Multimedia Security, IFIP TC6/TC11 Third Joint Working Conference, CMS '97, Athens, Greece, Chapman and Hall (1997) 215–226
3. Bhargava, B., Shi, C., Wang, Y.: MPEG video encryption algorithms. Multimedia Tools and Applications **24**(1) (2004) 57–79
4. Qiao, L., Nahrstedt, K.: Comparison of MPEG encryption algorithms. International Journal on Computers and Graphics (Special Issue on Data Security in Image Communication and Networks) **22**(3) (1998) 437–444
5. Uhl, A., Pommer, A.: Image and Video Encryption. From Digital Rights Management to Secured Personal Communication. Volume 15 of Advances in Information Security. Springer-Verlag (2005)
6. Stütz, T., Uhl, A.: Image confidentiality using Progressive JPEG. In: Proceedings of the International Conference on Information, Communications & Signal Processing, ICICS '05, Bangkok, Thailand (2005)
7. Fisch, M.M., Stögner, H., Uhl, A.: Layered encryption techniques for DCT-coded visual data. In: Proceedings (CD-ROM) of the European Signal Processing Conference, EUSIPCO '04, Vienna, Austria (2004) paper cr1361.
8. Wee, S., Apostolopoulos, J.: Secure scalable video streaming for wireless networks. In: Proceedings of the 2001 International Conference on Acoustics, Speech and Signal Processing (ICASSP 2001), Salt Lake City, Utah, USA (2001) invited paper.
9. Wee, S., Apostolopoulos, J.: Secure scalable streaming enabling transcoding without decryption. In: Proceedings of the IEEE International Conference on Image Processing (ICIP'01), Thessaloniki, Greece (2001)
10. Wee, S., Apostolopoulos, J.: Secure scalable streaming and secure transcoding with JPEG2000. In: Proceedings of the IEEE International Conference on Image Processing (ICIP'03). Volume I., Barcelona, Spain (2003) 547–551
11. Pennebaker, W., Mitchell, J.: JPEG – Still image compression standard. Van Nostrand Reinhold, New York (1993)
12. National Institute of Standards and Technology: FIPS-197 - advanced encryption standard (AES) (2001)
13. Mao, Y., Wu, M.: Security evaluation for communication-friendly encryption of multimedia. In: Proceedings of the IEEE International Conference on Image Processing (ICIP'04), Singapore, IEEE Signal Processing Society (2004)

Equivalence Analysis Among DIH, SPA, and RS Steganalysis Methods

Xiangyang Luo, Chunfang Yang, and Fenlin Liu

Information Engineering University, Zhengzhou, China
`xiangyangluo@126.com`

Abstract. steganography of images based on the use of the LSB (Least Significant Bit), SPA (Sample Pair Analysis), RS (Regular and Singular groups) method and DIH (Difference Image Histogram) method are three powerful steganalysis methods. A comparison analysis among DIH, SPA, and RS method is discussed, and a comparison of their proofs is presented. The process of proving includes three parts, and an equivalence relationship proposition is respectively proofed in every section. This proving offers a theory base for the study of an approach that can resist these three kinds of steganalysis methods synchronously.

1 Introduction

Steganography is one of the important research subjects in information security field. As a new art of covert communication, the main purpose of steganography is to convey messages secretly by concealing the very existence of messages under digital media files, such as images, audio, or video files. Similar to encryption and cryptanalysis, steganalysis attempts to defeat the goal of steganography. It is the art of detecting the existence of the secret message. Steganalysis finds applications in cyber warfare, computer forensics, tracking criminal activities over the Internet and gathering evidence for investigation. Steganalysis is also practiced for evaluating, identifying the weaknesses, and improving the security of steganographic systems.

Among the many steganographic methods involving images, LSB Steganography tools are now extremely widespread because of fine concealment, great capability of hidden message and easy realization. Making the detection of LSB steganography effective and reliable is a valuable topic for communication and multimedia security. Presently, there are some powerful LSB steganalysis methods, such as χ^2–statistical analysis[1], SPA method[2][3], RS steganalysis[4][5], DIH steganalysis[6][7][11] and so on. SPA steganalysis can detected the LSB steganography via sample pairs analysis. When the embedding ratio is more than 3%, it can estimate the embedding ratio with relatively high precision, and the average estimation error is 0.023. We improved SPA method, and proposed a more accurate LSB steganalysis method, called LSM (least square method) steganalysis in paper [8].

RS method, suitable for color and gray-scale images, is based on the number of the regular group and the singular one, and constructs a quadratic equation. The

H. Leitold and E. Markatos (Eds.): CMS 2006, LNCS 4237, pp. 161–172, 2006.
© IFIP International Federation for Information Processing 2006

embedding ratio of message in image is then estimated by solving the equation. This method can accurately estimate the length of the embedded messages when they are embedded randomly. An improved RS method based on dynamic masks is present in paper [9], which dynamically selects an appropriate mask for each image to reduce the initial deviation, and estimates the LSB embedding ratio more accurately. In addition, Andrew D. Ker [10] estimated the reliabilities of RS and SPA through a large number of experiments, and proposed some good improvement measures.

T. Zhang et al. [6][7][11] introduced a steganalytic method for detection of LSB embedding via different histograms of image, named DIH method. When the embedding ratio is more than 40% or less than 10%, the result is more accurate than that of RS method, and the speed of this method is faster.

In this paper, an equivalence analysis among DIH, SPA and RS method is discussed, and an equivalence proving of these three kinds of methods is presented. The proving process includes three parts, and three propositions are respectively proofed in these parts.

2   Principle of DIH, SPA, and RS Method

In this section, we simply describe the principle of DIH, SPA and RS method as a base of the equivalence proving.

2.1   Principle of DIH Method

A digital image can be represented by a set of pixels $s_1, s_2, \cdots, s_N$, where the index corresponds to the position of each pixel, and $\tilde{s}_k$ denotes the pixel adjacent to s_k (we consider adjacency in both dimensions, even though the indexes are not sequential). T. Zhang et al.[11] defines the pixel sets as follows:

$$H_n = \{s_k | s_k - \tilde{s}_k = n, k = 1, 2, \cdots, N, 0 \leq n \leq 255\} \tag{1}$$

$$G_{2m} = \{s_k | int\,(s_k/2) - int\,(\tilde{s}_k/2) = m, k = 1, 2, \cdots, N,\ 0 \leq m \leq 127\} \tag{2}$$

where $int\,(x)$ is the maximal integer that are not larger than x. Based on the relationship between G_{2m} and H_n, the following partition of G_{2m} can be obtained:

$$G_{2m} = A_{2m-1} \cup H_{2m} \cup B_{2m+1} \tag{3}$$

where

$$\begin{cases} A_{2m-1} = H_{2m-1} \cap G_{2m} = \{s_k | s_k \in G_{2m}, s_k \bmod 2 = 0, \tilde{s}_k \bmod 2 = 1, k = 1, \cdots, N\} \\ H_{2m} = H_{2m} \cap G_{2m} = \{s_k | s_k \in G_{2m}, (s_k \bmod 2) = (\tilde{s}_k \bmod 2), k = 1, \cdots, N\} \\ B_{2m+1} = H_{2m+1} \cap G_{2m} = \{s_k | s_k \in G_{2m}, s_k \bmod 2 = 1, \tilde{s}_k \bmod 2 = 0, k = 1, \cdots, N\} \end{cases} \tag{4}$$

Namely, for every s_k in A_{2m-1}, there is an adjacent pixel $\tilde{s}_k$, $s_k - \tilde{s}_k = 2m - 1$, and $int\,(s_k/2) - int\,(\tilde{s}_k/2) = m$; for every s_k in B_{2m+1}, there is an adjacent pixel $\tilde{s}_k$, $s_k - \tilde{s}_k = 2m + 1$, and $int\,(s_k/2) - int\,(\tilde{s}_k/2) = m$.

Define the transfer coefficient among the difference image histograms as follows:

$$a_{2m,2m-1} = \|A_{2m-1}\|/\|G_{2m}\|, \quad a_{2m,2m} = \|H_{2m}\|/\|G_{2m}\|,$$
$$a_{2m,2m+1} = \|B_{2m+1}\|/\|G_{2m}\| \tag{5}$$

where $\|\bullet\|$ denotes the cardinality of set $\bullet$. For $j = 0, \pm 1, 0 < a_{2m,2m+j} < 1$ or $a_{2m,2m+j} = 0$, and

$$a_{2m,2m-1} + a_{2m,2m} + a_{2m,2m+1} = 1. \tag{6}$$

DIH method denotes $h_m = \|H_m\|$, $g_{2m} = \|G_{2m}\|$ and f_m as the difference histograms of the detected image, the image in which after the LSB of every pixel is set as 0 and the image in which after the LSB of every pixel is flipped.

According to the definition of h_{2m+1}, it is known that h_{2m+1} comprises of $a_{2m,2m+1}g_{2m}$ and $a_{2m+2,2m+1}g_{2m+2}$. A majority of statistical tests show that for the natural images these two parts make an approximately equal contribution to h_{2m+1}, i.e.

$$a_{2m,2m+1}g_{2m} \approx a_{2m+2,2m+1}g_{2m+2}. \tag{7}$$

DIH method notes that $\alpha_m = a_{2m+2,2m+1}/a_{2m,2m+1}$, $\beta_m = a_{2m+2,2m+3}/a_{2m,2m-1}$ and $\gamma_m = g_{2m}/g_{2m+2}$, and makes the statistical hypothesis that satisfies

$$\alpha_m \approx \gamma_m, \tag{8}$$

For the natural image; but for the stego-images with LSB plane fully embedded, it satisfies

$$\alpha_m \approx 1. \tag{9}$$

Literature [6][11] selects the quadratic polynomial to simulate the relationship between α_m and p, and utilizes four key points $P_1 = (0, \gamma_m)$, $P_2 = (p, \alpha_m)$, $P_3 = (1, 1)$ and $P_4 = (2 - p, \beta_m)$ to obtain the estimation equation:

$$2d_1 p^2 + (d_3 - 4d_1 - d_2)p + 2d_2 = 0 \tag{10}$$

Where $d_1 = 1 - \gamma_m$, $d_2 = \alpha_m - \gamma_m$ and $d_3 = \beta_m - \gamma_m$. DIH regards the root of equation (10) whose absolute value is smaller as the estimate value of the embedding ratio p.

2.2 Principle of SPA Method

S. Dumitrescu et al.[2] denotes a pair of pixels as a two-tuple (s_i, s_j), $1 \leq i, j \leq N$, where N is the total number of pixels of an image. Then an estimation equation of the embedding ratio is based on the following important hypothesis:

$$E\{\|X_{2m+1}\|\} = E\{\|Y_{2m+1}\|\}, \tag{11}$$

where X_{2m+1} is the multiset consisting of the adjacent pixel pairs, for each (s_i, s_j) in X_{2m+1}, $|s_i - s_j| = 2m + 1$ and the even component in X_{2m+1} is larger; Y_{2m+1} is also the multiset consisting of the adjacent pixel pairs, for each (s_i, s_j) in Y_{2m+1}, $|s_i - s_j| = 2m + 1$ and the odd component in Y_{2m+1} is larger.

The other important multisets are defined in paper [2], such as C_m, D_n, where C_m is the multiset consisting of the adjacent pixel pairs whose values differ by m in the first $b - 1$ bits (b is the number of bits to represent each pixel value) (i.e., by right shifting one bit and then measuring the difference), and D_n is the multiset that consists of the adjacent pixel pairs whose values differ by n. The D_{2m+1} can be partitioned into two submultiset X_{2m+1} and Y_{2m+1}, and they satisfy $X_{2m+1} = D_{2m+1} \cap C_{m+1}$, $Y_{2m+1} = D_{2m+1} \cap C_m$, $0 \le m \le 2^{b-1} - 2$, and $X_{2^b-1} = \phi$, $Y_{2^b-1} = D_{2^b-1}$.

Considering the estimating precision, the literature [2] uses the hypothesis

$$E\left\{\left\|\bigcup_{m=i}^{j} X_{2m+1}\right\|\right\} = E\left\{\left\|\bigcup_{m=i}^{j} Y_{2m+1}\right\|\right\} \tag{12}$$

to replace (11), and then derives a more robust quadratic equations to estimate p.

2.3   Principle of RS Method

RS method partitions an image into $\lceil \frac{N}{n} \rceil$ groups of n adjacent pixels, where N is the total number of pixels in an image. In [5], the authors considered the case of $n = 4$. A discrimination function $f(\bullet)$ captures the smoothness of a group of pixels; and, we define three invertible operations $F_n(x)$, $n = -1, 0, 1$ on a pixel x, where F_1 and F_{-1} are applied to a group of pixel values through the mask M and $-M$. Mask M, an n-tuple with components 0 and 1, specifies where and how pixel values are to be modified; $-M$ is the n-tuple with the minus components of M, for example, if $M = (1, 0, 1, 0)$, then $-M = (-1, 0, -1, 0)$. Given a mask, operations F_1 and F_{-1}, and the discrimination function f, a pixel group G can be classified into one of the three categories described below:

$$G \in R(M) \Leftrightarrow f(F(G)) > f(G)$$

$$G \in S(M) \Leftrightarrow f(F(G)) < f(G)$$

$$G \in U(M) \Leftrightarrow f(F(G)) = f(G) \tag{13}$$

Where $R(M)$, $S(M)$ and $U(M)$ are respectively called Regular, Singular, and Unusable Groups. RS method is based on the statistical hypothesis that when no message is embedded in an image, the following equations hold:

$$E\{\|R(M)\|\} = E\{\|R(-M)\|\} \tag{14}$$

$$E\{\|S(M)\|\} = E\{\|S(-M)\|\}. \tag{15}$$

RS method builds a quadratic equation to estimate the embedding ratio p based on above-mentioned hypotheses (14) and (15), and the coefficients of the equation can be obtained by counting the number of Regular and Singular Groups with mask M and $-M$ in the examined image.

3   Comparison Among DIH, SPA and RS Method

In this section, the comparative analysis among DIH, SPA and RS method will be given to prove their equivalence.

3.1   Equivalence Between DIH and SPA Method

Proposition 1: The hypothesis (7) of DIH method is equivalent to the hypothesis (11) of SPA method.

Prove:
   From equation (5), the following equations can be obtained:

$$a_{2m,2m+1}g_{2m} = (\|B_{2m+1}\|/\|G_{2m}\|)\|G_{2m}\| = \|B_{2m+1}\|,$$

$$a_{2m+2,2m+1}g_{2m+2} = (\|A_{2m+1}\|/\|G_{2m+2}\|)\|G_{2m+2}\| = \|A_{2m+1}\|.$$

Thus, the hypothesis (7) of DIH method can be converted into

$$\|A_{2m+1}\| = \|B_{2m+1}\|. \tag{16}$$

From (1), we can denote H_n as a set of pixel s_k whose value is larger than that of an adjacent pixel $\tilde{s}_k$ by$n$. And D_n is a set of all pairs of adjacent pixels whose values differ by$n$. Thus, the adjacent pixels s_k and $\tilde{s}_k$ whose values differ by n are the elements of H_n and H_{-n} respectively, and the pixel pairs $(s_k, \tilde{s}_k)$ and $(\tilde{s}_k, s_k)$ are both the elements ofD_n. Therefore, the result of$H_n \cup H_{-n}$ is D_n.
   From (2), we can denote G_{2m} as a set of pixel s_k whose value is larger than that of an adjacent pixel $\tilde{s}_k$ by n in the first $b-1$ bits. And C_m is a set of all pairs of adjacent pixels whose values differ by n in the first $b-1$ bits. So, the above adjacent pixels s_k and $\tilde{s}_k$ are respectively the elements of G_{2m} andG_{-2m}, and the pixel pairs $(s_k, \tilde{s}_k)$ and $(\tilde{s}_k, s_k)$ are the elements ofC_m. Thus, $G_{2m} \cup G_{-2m}$equalsC_m.
   From (4), it follows: denote A_{2m+1} as a set of pixel s_k whose value is larger than that of an adjacent pixel $\tilde{s}_k$ by $2m+1$ and in the first $b-1$ bits s_k is larger than $\tilde{s}_k$ by$m+1$. And X_{2m+1} is a set of all pairs of adjacent pixels whose values differ by $2m+1$ and $m+1$ in the first $b-1$ bits. Hence, the above adjacent pixels s_k and $\tilde{s}_k$ are the elements of A_{2m+1} and A_{-2m-1} respectively, and the pairs $(s_k, \tilde{s}_k)$ and $(\tilde{s}_k, s_k)$ are the elements of X_{2m+1}. Thereby, $A_{2m+1} \cup A_{-2m-1}$ is equivalent toX_{2m+1}.
   As above, it follows: denote B_{2m+1} as a set of pixel s_k whose value is larger than that of an adjacent pixel $\tilde{s}_k$ by $2m+1$ and s_k is larger than $\tilde{s}_k$ by m in the first $b-1$ bits. And Y_{2m+1} is a set of all pairs of adjacent pixels whose values differ by $2m+1$ and m in the first $b-1$ bits. Therefore, the above adjacent pixels s_k and $\tilde{s}_k$ are the elements of B_{2m+1} and B_{-2m-1} respectively, and the pixel pairs $(s_k, \tilde{s}_k)$ and $(\tilde{s}_k, s_k)$ are the elements ofY_{2m+1}. Thus, $B_{2m+1} \cup B_{-2m-1}$ is equivalent to Y_{2m+1}.
   If an arbitrarys_k belongs to H_n, G_{2m}, A_{2m+1} or B_{2m+1}, there must be a corresponding adjacent element $\tilde{s}_k$ belonging to H_{-n}, G_{-2m}, A_{-2m-1} or B_{-2m-1} respectively and vice versa. Consequently, it is held that

$$\|H_n\| = \|H_{-n}\|, \quad \|G_{2m}\| = \|G_{-2m}\|,$$
$$\|A_{2m+1}\| = \|A_{-2m-1}\|, \quad \|B_{2m+1}\| = \|B_{-2m-1}\|,$$

That is,

$$\|H_n\| = \frac{1}{2}\left(\|H_n\| + \|H_{-n}\|\right), \quad \|G_{2m}\| = \frac{1}{2}\left(\|G_{2m}\| + \|G_{-2m}\|\right),$$

$$\|A_{2m+1}\| = \frac{1}{2}\left(\|A_{2m+1}\| + \|A_{-2m-1}\|\right), \quad \|B_{2m+1}\| = \frac{1}{2}\left(\|B_{2m+1}\| + \|B_{-2m-1}\|\right). \tag{17}$$

In brief, DIH and SPA method adopt different means to build estimation equations: DIH method utilizes the similarity degree $\alpha_m = a_{2m+2,2m+1}/a_{2m,2m+1}$ between A_{2m+1}/B_{2m+1} and g_{2m}/g_{2m+2}, the ratio between $a_{2m,2m+1}g_{2m}$ and $a_{2m+2,2m+1}g_{2m+2}$ in h_{2m+1}, to model the relationship between α_m and p; and SPA method constructs the estimation equation of p through the transform probability among states that the adjacent pixel pairs belong to before and after embedding. However, the assumption (7) of DIH method is equivalent to the assumption $E\{\|Y_{2m+1}\|\} = E\{\|X_{2m+1}\|\}$ of SPA method in nature. In fact, both of them are based on the same hypothesis: for an natural image, in the adjacent pixels differing by $2m + 1$, their probabilities differing by m or $m + 1$ are equal. Consequently, The combination of (7) in $m = 0, \cdots, j, \sum_{m=i}^{j} a_{2m,2m+1}g_{2m} = \sum_{m=i}^{j} a_{2m+2,2m+1}g_{2m+2}$, namely $\left\|\bigcup_{m=i}^{j} A_{2m+1}\right\| = \left\|\bigcup_{m=i}^{j} B_{2m+1}\right\|$, is equivalent to the assumption (12) of SPA method.

3.2 Equivalence Between DIH and RS Method

Proposition 2: When $n = 2, m = 0, \cdots, 2^{b-1} - 2$, the hypotheses (14) and (15) of RS method is equivalent to the combination of hypothesis (7) of DIH method.

Prove:

When $n = 2$, M can be one of the four cases: $(1, 0), (0, 1), \begin{pmatrix} 1 \\ 0 \end{pmatrix}$ and $\begin{pmatrix} 0 \\ 1 \end{pmatrix}$.

Consider the case $M = (1, 0)$, the process of prove is as follows.

When $M = (1, 0)$, the pixels s_k and $\tilde{s}_k$ are horizontally adjacent. Two pixel-sets $H_{00,2m}$ and $H_{11,2m}$ are defined here, both of which s_k is larger than $\tilde{s}_k$ by $2m$. Furthermore, the LSBs of s_k in $H_{00,2m}$ and its adjacent pixel $\tilde{s}_k$ are both zeros and the LSBs of s_k in $H_{11,2m}$ and its adjacent pixel $\tilde{s}_k$ are both ones. Applying $M = (1, 0)$ into the detecting image, all the horizontal adjacent pixel pairs in the image can be partitioned by two means: $R(M) + S(M)$ and $R(-M) + S(-M)$.

1. If the adjacent pixel pair $(s_k, \tilde{s}_k)$ belongs to $R(M)$, then s_k and $\tilde{s}_k$ are equivalent, or the larger s_k becomes more larger, or the smaller s_k becomes more smaller through applying the operation F_1 into s_k. Namely, $(s_k, \tilde{s}_k)$ may be under one of the below cases:

$$\begin{cases} s_k > \tilde{s}_k, \text{ if } s_k \bmod 2 = 0 \\ s_k < \tilde{s}_k, \text{ if } s_k \bmod 2 = 1 \\ s_k = \tilde{s}_k \end{cases}.$$

Therefore, s_k belongs to

$$\left(\bigcup_{m=1}^{2^{b-1}-1} A_{2m-1}\right) \cup \left(\bigcup_{m=1}^{2^{b-1}-1} A_{-2m+1}\right) \cup \left(\bigcup_{m=0}^{2^{b-1}-1} H_{00,2m}\right) \cup \left(\bigcup_{m=0}^{2^{b-1}-1} H_{11,-2m}\right).$$

Replacing m by $m+1$ in the above formula, it follows that

$$\left(\bigcup_{m=0}^{2^{b-1}-2} A_{2m+1}\right) \cup \left(\bigcup_{m=0}^{2^{b-1}-2} A_{-2m-1}\right) \cup \left(\bigcup_{m=-1}^{2^{b-1}-2} H_{00,2m+2}\right) \cup \left(\bigcup_{m=-1}^{2^{b-1}-2} H_{11,-2m-2}\right),$$

Namely,

$$\left(\bigcup_{m=0}^{2^{b-1}-2} A_{2m+1}\right) \cup \left(\bigcup_{m=0}^{2^{b-1}-2} A_{-2m-1}\right) \cup H_{00,0} \cup H_{11,0} \cup \left(\bigcup_{m=0}^{2^{b-1}-2} H_{00,2m+2}\right) \cup$$

$$\cup \left(\bigcup_{m=0}^{2^{b-1}-2} H_{11,-2m-2}\right). \tag{18}$$

If s_k belongs to $\left(\bigcup_{m=0}^{2^{b-1}-2} A_{2m+1}\right) \cup \left(\bigcup_{m=0}^{2^{b-1}-2} A_{-2m-1}\right) \cup H_{00,0} \cup H_{11,0}$, then its

adjacent pixel $\tilde{s}_k$ must also belong to $\left(\bigcup_{m=0}^{2^{b-1}-2} A_{2m+1}\right) \cup \left(\bigcup_{m=0}^{2^{b-1}-2} A_{-2m-1}\right) \cup$

$H_{00,0} \cup H_{11,0}$ and $(s_k, \tilde{s}_k)$ must be of $R(M)$. If s_k belongs to $\left(\bigcup_{m=0}^{2^{b-1}-2} H_{00,2m+2}\right) \cup$

$\left(\bigcup_{m=0}^{2^{b-1}-2} H_{11,-2m-2}\right)$, then $(s_k, \tilde{s}_k)$ must be of $R(M)$, but $\tilde{s}_k$ must not be an

element of $\left(\bigcup_{m=0}^{2^{b-1}-2} H_{00,2m+2}\right) \cup \left(\bigcup_{m=0}^{2^{b-1}-2} H_{11,-2m-2}\right)$. Therefore,

$$\|R(M)\| = \frac{1}{2}\left(\left\|\bigcup_{m=0}^{2^{b-1}-2} A_{2m+1}\right\| + \left\|\bigcup_{m=0}^{2^{b-1}-2} A_{-2m-1}\right\| + \|H_{00,0}\| + \|H_{11,0}\|\right)$$

$$+ \left\|\bigcup_{m=0}^{2^{b-1}-2} H_{00,2m+2}\right\| + \left\|\bigcup_{m=0}^{2^{b-1}-2} H_{11,-2m-2}\right\|$$

$$\tag{19}$$

1. If $(s_k, \tilde{s}_k)$ belongs to $R(-M)$, then s_k and $\tilde{s}_k$ are equivalent, or the larger s_k becomes smaller or the smaller s_k becomes larger through applying F_1 into s_k. Namely, $(s_k, \tilde{s}_k)$ may be under one of the following cases:

$$\begin{cases} s_k > \tilde{s}_k, \text{ if } s_k \bmod 2 = 1 \\ s_k < \tilde{s}_k, \text{ if } s_k \bmod 2 = 0 \\ s_k = \tilde{s}_k \end{cases},$$

Similar to i), we can obtain

$$
\|R(-M)\| = \tfrac{1}{2} \left(\left\| \bigcup_{m=0}^{2^{b-1}-1} B_{2m+1} \right\| + \left\| \bigcup_{m=0}^{2^{b-1}-1} B_{-2m-1} \right\| + \|H_{00,0}\| + \|H_{11,0}\| \right) .
$$
$$
+ \left\| \bigcup_{m=1}^{2^{b-1}-1} H_{11,2m} \right\| + \left\| \bigcup_{m=1}^{2^{b-1}-1} H_{00,2m} \right\|
$$

(20)

The proving process will be specified in Appendix.

1. If $(s_k, \tilde{s}_k)$ belongs to $S(M)$, then $(s_k, \tilde{s}_k)$ may be classified into two categories:

$$
\begin{cases} s_k > \tilde{s}_k, & \text{if } s_k \bmod 2 = 1 \\ s_k < \tilde{s}_k, & \text{if } s_k \bmod 2 = 0 \end{cases} .
$$

Similar to i) and ii), we can prove

$$
\|S(M)\| = \frac{1}{2} \left(\left\| \bigcup_{m=0}^{2^{b-1}-1} B_{2m+1} \right\| + \left\| \bigcup_{m=0}^{2^{b-1}-1} B_{-2m-1} \right\| \right) +
$$
$$
\left\| \bigcup_{m=0}^{2^{b-1}-2} H_{11,2m+2} \right\| + \left\| \bigcup_{m=0}^{2^{b-1}-2} H_{00,-2m-2} \right\| .
$$

(21)

1. If $(s_k, \tilde{s}_k)$ belongs to $S(-M)$, $(s_k, \tilde{s}_k)$ may be classified into two classes:

$$
\begin{cases} s_k > \tilde{s}_k, & \text{if } s_k \bmod 2 = 0 \\ s_k < \tilde{s}_k, & \text{if } s_k \bmod 2 = 1 \end{cases} .
$$

Similar to i), we can obtain that

$$
\|S(-M)\| = \frac{1}{2} \left(\left\| \bigcup_{m=0}^{2^{b-1}-2} A_{2m+1} \right\| + \left\| \bigcup_{m=0}^{2^{b-1}-2} A_{-2m-1} \right\| \right)
$$
$$
+ \left\| \bigcup_{m=0}^{2^{b-1}-2} H_{00,2m+2} \right\| + \left\| \bigcup_{m=0}^{2^{b-1}-2} H_{11,-2m-2} \right\| .
$$

(22)

From the definitions of A_{2m+1}, B_{2m+1}, $H_{00,2m}$ and $H_{11,2m}$, it can be shown that: if a arbitrary pixel s_k belongs to $\bigcup_{m=0}^{2^{b-1}-2} A_{2m+1}$, $\bigcup_{m=0}^{2^{b-1}-1} B_{2m+1}$, $\bigcup_{m=0}^{2^{b-1}-2} H_{00,2m+2}$ or $\bigcup_{m=0}^{2^{b-1}-2} H_{11,2m+2}$, then there must be only one adjacent $\tilde{s}_k$ belonging to $\bigcup_{m=0}^{2^{b-1}-2} A_{-2m-1}$, $\bigcup_{m=0}^{2^{b-1}-1} B_{-2m-1}$, $\bigcup_{m=0}^{2^{b-1}-2} H_{00,-2m-2}$ or $\bigcup_{m=0}^{2^{b-1}-2} H_{11,-2m-2}$. Hence

$$
\left\| \bigcup_{m=0}^{2^{b-1}-2} A_{2m+1} \right\| = \left\| \bigcup_{m=0}^{2^{b-1}-2} A_{-2m-1} \right\| , \quad \left\| \bigcup_{m=0}^{2^{b-1}-1} B_{2m+1} \right\| = \left\| \bigcup_{m=0}^{2^{b-1}-1} B_{-2m-1} \right\| ,
$$

$$\left\|\bigcup_{m=0}^{2^{b-1}-2} H_{00,2m+2}\right\| = \left\|\bigcup_{m=0}^{2^{b-1}-2} H_{00,-2m-2}\right\|,$$

$$\left\|\bigcup_{m=0}^{2^{b-1}-2} H_{11,2m+2}\right\| = \left\|\bigcup_{m=0}^{2^{b-1}-2} H_{11,-2m-2}\right\|. \tag{23}$$

Based on hypotheses (14) and (15), we can obtain

$$\frac{1}{2}\left(\left\|\bigcup_{m=0}^{2^{b-1}-2} A_{2m+1}\right\| + \left\|\bigcup_{m=0}^{2^{b-1}-2} A_{-2m-1}\right\| + \|H_{00,0}\| + \|H_{11,0}\|\right)$$
$$+ \left\|\bigcup_{m=0}^{2^{b-1}-2} H_{00,2m+2}\right\| + \left\|\bigcup_{m=0}^{2^{b-1}-2} H_{11,-2m-2}\right\|$$
$$= \frac{1}{2}\left(\left\|\bigcup_{m=0}^{2^{b-1}-1} B_{2m+1}\right\| + \left\|\bigcup_{m=0}^{2^{b-1}-1} B_{-2m-1}\right\| + \|H_{00,0}\| + \|H_{11,0}\|\right), \tag{24}$$
$$+ \left\|\bigcup_{m=0}^{2^{b-1}-2} H_{11,2m+2}\right\| + \left\|\bigcup_{m=0}^{2^{b-1}-2} H_{00,-2m-2}\right\|$$

and

$$\frac{1}{2}\left(\left\|\bigcup_{m=0}^{2^{b-1}-1} B_{2m+1}\right\| + \left\|\bigcup_{m=0}^{2^{b-1}-1} B_{-2m-1}\right\|\right) + \left\|\bigcup_{m=0}^{2^{b-1}-2} H_{11,2m+2}\right\| +$$
$$\left\|\bigcup_{m=0}^{2^{b-1}-2} H_{00,-2m-2}\right\| = \frac{1}{2}\left(\left\|\bigcup_{m=0}^{2^{b-1}-2} A_{2m+1}\right\| + \left\|\bigcup_{m=0}^{2^{b-1}-2} A_{-2m-1}\right\|\right) + . \tag{25}$$
$$\left\|\bigcup_{m=0}^{2^{b-1}-2} H_{00,2m+2}\right\| + \left\|\bigcup_{m=0}^{2^{b-1}-2} H_{11,-2m-2}\right\|$$

From (23), (24) and (25), it can be further obtained that

$$\left\|\bigcup_{m=0}^{2^{b-1}-2} A_{2m+1}\right\| = \left\|\bigcup_{m=0}^{2^{b-1}-1} B_{2m+1}\right\|. \tag{26}$$

Usually, when $b = 8$, the probability of two adjacent pixels differing by 255, viz.$2^b - 1$, is nearly zero. As a result, (26) can be shown in the following way

$$\left\|\bigcup_{m=0}^{2^{b-1}-2} A_{2m+1}\right\| = \left\|\bigcup_{m=0}^{2^{b-1}-2} B_{2m+1}\right\|. \tag{27}$$

Accordingly, when $M = (1,0)$, the assumptions (14) and (15) of RS method are equivalent to the combination of assumption (7) of DIH method.

In the same way, the equivalence relationship can be proofed when $M = (0,1)$, $\binom{1}{0}$ or $\binom{0}{1}$.

To sum up, when $n = 2$, $m = 0$, $\cdots$, $2^{b-1} - 2$, the assumptions (14) and (15) of RS method are equivalent to the combination of assumption (7) of DIH method.

3.3   Equivalence Between RS and SPA Method

Proposition 3: When $n = 2$, the hypotheses (14) and (15) of RS method are equivalent to the combination hypothesis

$$E\left\{\left\|\bigcup_{m=0}^{2^{b-1}-2} X_{2m+1}\right\|\right\} = E\left\{\left\|\bigcup_{m=0}^{2^{b-1}-2} Y_{2m+1}\right\|\right\} \text{ of SPA method.}$$

Prove:

In section 3.1, the hypothesis (7) of DIH method is equivalent to the hypothesis (11) of SPA method. And in section 3.2, when $n = 2$, $m = 0$, $\cdots$, $2^{b-1} - 2$, the assumptions (14) and (15) of RS method are equivalent to the combination of assumption (7) of DIH method. Hence, when $n = 2$, the assumptions (14) and (15) of RS method are equivalent to the combination hypothesis

$$E\left\{\left\|\bigcup_{m=0}^{2^{b-1}-2} X_{2m+1}\right\|\right\} = E\left\{\left\|\bigcup_{m=0}^{2^{b-1}-2} Y_{2m+1}\right\|\right\} \text{ of SPA method.}$$

From 3.1, 3.2 and 3.3, it can be found that all of three methods depend on the weak correlation between the LSB plane and the remained bit planes though the different implementation methods. This weak correlation decreases with the increase of the embedded message and is represented as the assumption (12). From this section, the assumptions (7) and $\sum_{m=i}^{j} a_{2m,2m+1} g_{2m} = \sum_{m=i}^{j} a_{2m+2,2m+1} g_{2m+2}$ in DIH method are equivalent to (11) and (12) in SPA method; when $n = 2$, the assumptions (14) and (15) based on RS method equal to the special example of (12), viz. $\left\|\bigcup_{m=0}^{2^{b-1}-2} A_{2m+1}\right\| = \left\|\bigcup_{m=0}^{2^{b-1}-2} B_{2m+1}\right\|$. Consequently, it is concluded that DIH, SPA and RS methods are based on the same kind of hypothesis and are virtually similar.

4   Conclusions

Image steganalysis has attracted the increasing attention recently, and the LSB steganalysis is one of the most active research topics. SPA, RS and DIH are three powerful LSB steganalysis methods. In this paper, we make a comparison analysis among SPA, RS and DIH method, and present an equivalence proving of them. The proving process includes three parts, and three propositions are respectively proofed in these sections. This equivalence proving offers a theory base for the study of an approach that can synchronously resist these three kinds of steganalysis methods, which we will aim at.

References

1. A. Westfeld, "Detecting low embedding rates", Proceedings of the 5th Information Hiding Workshop, Volume 2578 of Springer LNCS, 2003, 324–339.
2. Sorina Dumitrescu, Xiaolin Wu, and Zhe Wang, "Detection of LSB Steganography via Sample Pair Analysis", IEEE Transactions on Signal Processing, VOL.51, NO.7, 2003, 1995–2007.

3. S. Dumitrescu, X. Wu, and Z. Wang, "Detection of LSB Steganography via Sample Pair Analysis", Proceedings of the 5th Information Hiding Workshop, Volume 2578 of Springer LNCS, 2003, 355–372.
4. J. Fridrich, M. Goljan, "Practical Steganalysis of Digital Images – State of the Art", In Delp III, E.J., Wong, P.W., eds.: Security and Watermarking of Multimedia Contents IV. Volume 4675 of SPIE, 2002, 1-13.
5. J. Fridrich, M. Goljan, and R.Du, "Reliable detection of LSB steganography in color and grayscale images," Proceedings of ACM Workshop Multimedia Security, Ottawa, ON, Canada, Oct.5, 2001, 27-30.
6. T. Zhang; X. Ping. "Reliable detection of LSB steganography based on the difference image histogram", Proceedings of IEEE ICASSP 2003, Volume 3: 545-548.
7. T. Zhang, X. Ping. "A new approach to reliable detection of LSB steganography in natural images", Signal Processing, Elsevier, Volume 83 ,Issue 10, 2003, 2085-2093.
8. P. Lu, X. Luo et. al. "An Improved Sample Pairs Method for Detection of LSB Embedding", Proceedings of the 6th Information Hiding Workshop, Volume 3200 of Springer LNCS, 2004, 116-128.
9. X. Luo, B. Liu, F. Liu, "Improved RS Method for Detection of LSB Steganography", Proceedings of International Conference on Computational Sciences & Its Applications (ICCSA 2005), Volume 3481 of Springer LNCS, 2005, 508-516.
10. Andrew D. Ker, "Steganalysis of LSB Matching in Grayscale Images", IEEE Signal Processing Letters, vol. 12, No. 6, June, 2005, 441-444.
11. T. Zhang, X. Ping. "Reliable Detection of Spatial LSB SteganOgraphy Based on Difference Histogram", Journal of Software, 2004, 15(1)151~158. (Chinese).

Appendix

If $(s_k, \tilde{s}_k)$ belongs to $R(-M)$, then s_k and $\tilde{s}_k$ are equivalent, or the larger s_k becomes smaller or the smaller s_k becomes larger through applying F_1 into s_k. Namely, $(s_k, \tilde{s}_k)$ may be under one of the following three cases:

$$\begin{cases} s_k > \tilde{s}_k, \text{ if } s_k \bmod 2 = 1 \\ s_k < \tilde{s}_k, \text{ if } s_k \bmod 2 = 0 \ , \\ s_k = \tilde{s}_k \end{cases}$$

Then, s_k belongs to

$$\left(\bigcup_{m=0}^{2^{b-1}-1} B_{2m+1} \right) \cup \left(\bigcup_{m=0}^{2^{b-1}-1} B_{-2m-1} \right) \cup \left(\bigcup_{m=0}^{2^{b-1}-1} H_{11,2m} \right) \cup \left(\bigcup_{m=0}^{2^{b-1}-1} H_{00,-2m} \right).$$

Replacing m by $m+1$ in $\left(\bigcup_{m=0}^{2^{b-1}-1} H_{11,2m} \right) \cup \left(\bigcup_{m=0}^{2^{b-1}-1} H_{00,-2m} \right)$ of the above formula, the below formula can be obtained:

$$\left(\bigcup_{m=0}^{2^{b-1}-1} B_{2m+1} \right) \cup \left(\bigcup_{m=0}^{2^{b-1}-1} B_{-2m-1} \right) \cup \left(\bigcup_{m=-1}^{2^{b-1}-2} H_{11,2m+2} \right) \cup \left(\bigcup_{m=-1}^{2^{b-1}-2} H_{00,-2m-2} \right),$$

viz.

$$\left(\bigcup_{m=0}^{2^{b-1}-1} B_{2m+1} \right) \cup \left(\bigcup_{m=0}^{2^{b-1}-1} B_{-2m-1} \right) \cup H_{00,0} \cup H_{11,0} \cup \left(\bigcup_{m=0}^{2^{b-1}-2} H_{11,2m+2} \right)$$
$$\cup \left(\bigcup_{m=0}^{2^{b-1}-2} H_{00,-2m-2} \right). \tag{28}$$

If s_k belongs to $\left(\bigcup_{m=0}^{2^{b-1}-1} B_{2m+1} \right) \cup \left(\bigcup_{m=0}^{2^{b-1}-1} B_{-2m-1} \right) \cup H_{00,0} \cup H_{11,0}$, then its adjacent pixel $\tilde{s}_k$ must also be one element of

$\left(\bigcup_{m=0}^{2^{b-1}-1} B_{2m+1} \right) \cup \left(\bigcup_{m=0}^{2^{b-1}-1} B_{-2m-1} \right) \cup H_{00,0} \cup H_{11,0}$ and the pixel pair $(s_k, \tilde{s}_k)$

must be the element of $R(-M)$. If s_k belongs to

$\left(\bigcup_{m=0}^{2^{b-1}-2} H_{11,2m+2} \right) \cup \left(\bigcup_{m=0}^{2^{b-1}-2} H_{00,-2m-2} \right)$, then $(s_k, \tilde{s}_k)$ must belong to $R(-M)$,

but $\tilde{s}_k$ must not be the element of

$\left(\bigcup_{m=0}^{2^{b-1}-2} H_{11,2m+2} \right) \cup \left(\bigcup_{m=0}^{2^{b-1}-2} H_{00,-2m-2} \right)$. Therefore, we can obtain the equation

(20),

$$\|R(-M)\| = \frac{1}{2} \left(\left\| \bigcup_{m=0}^{2^{b-1}-1} B_{2m+1} \right\| + \left\| \bigcup_{m=0}^{2^{b-1}-1} B_{-2m-1} \right\| + \|H_{00,0}\| + \|H_{11,0}\| \right.$$
$$\left. + \left\| \bigcup_{m=1}^{2^{b-1}-1} H_{11,2m} \right\| + \left\| \bigcup_{m=1}^{2^{b-1}-1} H_{00,2m} \right\| \right).$$

A Flexible and Open DRM Framework

Kristof Verslype and Bart De Decker

Department of Computer Science, K.U.Leuven,
Celestijnenlaan 200A, B-3001 Leuven, Belgium
{kristof.verslype, bart.dedecker}@cs.kuleuven.be

Abstract. Current DRM implementations rely on obfuscating the inner working of the DRM client. Moreover, the rights to consume content are rather device bound than person bound. We present a first step towards an open DRM framework which is based on the security of its building blocks. The presented framework binds the right to consume content to persons instead of to devices. An extension of the current TPM specification is proposed to enhance the security of DRM clients.

Keywords: DRM, Digital Rights Management, TPM, smart card, e-commerce.

1 Introduction

DRM (Digital Rights Managent) is a technique to allow owners of digital content to control access to and distribution of this content and to restrict its usage in various ways which can be specified by the owner or his/her delegates.

Current implementations are closed source and most details are hidden, because, currently, the security of the DRM technologies relies on the secrecy of algorithms in the DRM client. Due to this approach, even the DRM technologies that are considered as the most mature and most secure are broken (see [7]).

On the other hand, current technologies lack flexibility in many ways. This paper is intended to be a next step towards more flexible DRM. Firstly, the ability to consume content is bound to the consumers themselves, where currently, this is bound to one or more devices. Secondly, the presented framework is flexible in the sense that it is based on building blocks that can be replaced if they are no longer considered appropriate. The required building blocks are not all equally mature, but we can expect that this will impove in the near future.

The basis of the person binding solution is to have two types of licenses: a content license and a root license. A content license gives a specific consumer the right to perform some actions on DRM protected content. A root license is device bound and enables the consumer to use all his content licenses on that device. A reasonable extension of the current TPM specification is proposed in order to obtain a DRM framework that is hard to break.

In the next section, the general DRM concepts are introduced. In section 3, some building blocks are briefly explained. Section 4 presents the DRM framework.

H. Leitold and E. Markatos (Eds.): CMS 2006, LNCS 4237, pp. 173–184, 2006.

Section 5 analyses the security properties. Section 6 compares the construction with existing implementations. The paper ends with the conclusions and future work in section 7.

2   General DRM Concepts

This section introduces the main concepts of DRM. An introduction to the technical aspects of DRM can be found in [6].

Content refers to the data to protect. This can be multimedia, text, applications or other data constructs. Performing an action on content is called content *consumption*. The *producer* is the entity that owns the rights to distribute and sell content. The *consumer* obtains and consumes content and the *publisher* owns and manages the DRM system used to distribute content. The online *DRM system* is the (set of) server(s) offering DRM services to consumers and producers. The *DRM client* is the entity at the consumer side that is responsible for securely performing the DRM-specific operations such as the enforcement of the usage rules (the rights).

Usually, a *prevention mechanism* (encryption) is combined with a *detection mechanism*, which allows for identifying the source of misuse when illegally distributed content is found. Usage rules are associated with the corresponding DRM protected content. *Licenses* introduce a separation of DRM protected content and the usage rules associated with it. The latter describe the actions allowed on the content. Before consumers are able to use protected content, they first have to obtain a corresponding license that enables them to use the content according to the usage rules described therein.

A *contract* can be agreed between publisher and producer stating the terms of the agreement (e.g. royalties and usage rules of the licenses).

3   Cryptographic Primitives and Techniques

Besides the classical cryptographic primitives such as hash functions, digital signatures, certificates, public and private encryption, some less commonly known techniques will be needed to develop the DRM framework.

Watermarking (see [2]) embeds some information into content without noticeably changing the content. The watermark has to be undetectable by human perception. Strong watermarks survive manipulations such as D/A A/D conversions. It is one of the least mature technologies used in DRM.

White-box cryptography (see [3], [4]) assumes that the adversary can fully analyse the software implementation and the run-time instances, including the execution of cryptographic functions. White-box cryptography embeds a key in code such that it remains hidden from adversaries.

Code guarding (see [5], [1]) is a collection of techniques used to prevent tampering with the code during execution. Code guarding is useful when applications run on untrusted hosts.

TPM (Trusted Platform Module, see [12]) was specified by the Trusted Computing Group. The TPM guards the system security status during startup and runtime. This status can also be interrogated later on. The TPM has some private keys embedded that cannot be read by anyone. These all have their corresponding public keys and certificates. The TPM offers the possibility to protect key material for outsiders, to authenticate the system to third parties, to prove the system's security status to third parties, to generate random numbers, to seal content and to detect configuration changes. On top of the TPM, a Trusted environment is built and an API is offered to the applications.

The TPM has its own cryptographic co-processor which cannot be addressed directly from outside the TPM. The TPM has volatile and non-volatile memory which can be used to store information in the TPM. Authentication to the TPM is based on knowledge of a shared secret. Other entities can get authorized by the TPM owner to access the TPM. It is possible to establish a confidentiality protecting transport session between (remote or local) processes and the TPM with the consent of the TPM owner. The TPM v1.2 has its own mechanism to do access control of software processes to the TPM. The TPM has a symmetric encryption engine to protect the confidentiality in transport sessions, to encrypt data that is stored outside the TPM, ... The TPM can store data such as keys in an encrypted form on the hard disk, such that it can only be decrypted by that TPM. The data is also integrity protected. The TPM v1.2 specifications offers monotonic counters. A process can get exclusive access to a monotonic counter such that it can be read out and incremented by that process. The TPM's Tick Counter can, with some external support enable secure time stamping. The TPM specification describes more functionality, but these are the most crucial ones for the DRM framework.

4   Framework

In this section, the framework is elaborated. We first divide the online DRM system into components. We then give a high level overview of our approach, followed by the assumptions made. Finally, the framework itself is described.

4.1   Components

In [8] different components in a generic DRM architecture were identified. We slightly adapt these to our specific needs. The result is shown in figure 1. In the center, we have the online DRM System, which consists of multiple services that are available to the consumer (via the DRM client) or the producer (via a producer tool). The *Import Service (IS)* is used to add new content to the online DRM system. The *Content Service (CS)* and *License Service (LS)* provide content and corresponding licenses. The *Registration Service (RS)* enables or disables a device to consume the consumer's content. The *Identification Service (IdS)* identifies the source in case illegally distributed content is found. The Client Setup Service is contacted to install or update a DRM client. The black

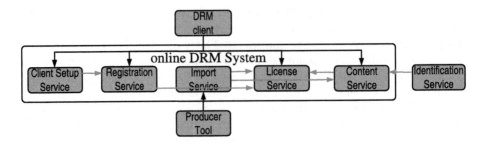

Fig. 1. Relevant DRM services

arrows indicate communication initiated by (and with the consent of) the consumer or producer. The grey arrows indicate communication between the online services.

4.2   Content License and Root License

We distinguish between content licenses and root licenses as shown in figure 2.

Content licenses are enablers for a consumer to consume specific content according to some predefined set of rights. Content Licenses are not bound to and thus independent of any device. A content license can be written formally as $("content",$ $Id_{license}, Id_{content}, Id_C, Id_{LS}, K^E_{content}, rights, date_{expiry})sig_{LS}$. *"content"* determines the license category, $Id_{license}$ uniquely identifies the license, $Id_{content}$ the content to which the rights apply, Id_C the license owner (the consumer) and Id_{LS} the issuing License Service; *rights* specify the associated rights. The encrypted content key $K^E_{content}$ can only be decrypted at the consumer's side with the help of a root license. A content license expires at $date_{expiry}$. The content license is signed by the issuing LS. We will often drop the word "content" as prefix when talking about content licenses.

Root licenses are enablers for a consumer to use content licenses on a specific device. Without a root license, a consumer cannot consume content on that device. Each user needs a separate root license for each device he/she wants to use. A root license is issued by a Registration Service and can be formally written as $("root", Id_C, Id_{RS}, Id_{lroot}, SK^E_{lroot}, date_{expiry})sig_{RS}$ which is analog to the content license. Each (registered) consumer has been assigned (but not given) one secret root key SK_{lroot}. It is encrypted in the root license of the consumer for that device such that only the TPM can get hold of it (even not the consumer himself). This key is required to decrypt the encrypted content key in content licenses.

We distinguish between public computers, semi-public computers and private devices. Public computers are used by a lot of (often occasional) users. These are for instance computers in a cybercafe or public library. Semi-public computers are used by a limited set af users. A typical example is the home computer used by all the family members. Private devices are typically used by a single person (its owner) examples are mobile phones and MP3 players. On public (and semi-public)

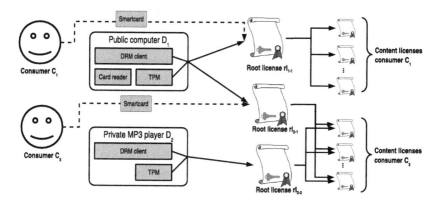

Fig. 2. A root license is required to consume DRM protected content on a specific device. A smartcard is required to consume content on (semi) public computers.

computers, we must avoid that anyone using that computer can consume the rights of another consumer. Therefore, the consumer need to authenticate to the DRM client. A simple login/password is not sufficient, because this can be shared by multiple persons. We thus need an authentication mechanism such that the DRM client is sure that the authenticating person is indeed the owner of the root license. Therefore, smartcards could be used: e.g. electronic identity cards, which are being issued by the governments of several countries. These electronic identity cards have the advantage that the owner is not keen to give or lend it to others and that each civilian has one. On the other hand, on private devices this smartcard authentication is not really necessary and would even be highly impractical. Smart card authentication is indicated with a dashed line in figure 2.

4.3   Assumptions

The different entities involved need certificates for authentication purposes, including the output devices such as display devices, which have to prove their trustworthiness. We assume that the output devices have decryption capabilities, and that the decrypted content in the output devices cannot be captured, such that the signal only travels the expected way.

The (new) methods described below must be added to the current TPM interface. The entity calling this functions must be authenticated to the TPM.

- A method $true/false \leftarrow storeKey(K_{name}, name, policy)$, which stores the key K_{name} into the non-volatile memory of the TPM. A policy $policy$ is associated with it, determining what the caller of the function or other entities are able to do with the key. The caller of this function becomes the owner of the TPM stored key K_{name}. The TPM also stores the name of the key and the identity of the key owner. The latter can be an application, but still, the TPM owner must give his/her consent before this method is executed to prevent abuse.

- A method $Id_{K_{name}} \leftarrow loadKey(name)$, which loads a TPM stored key with a given name in the volatile memory of the TPM. The TPM checks the policy associated with the key before the function is executed. i.e. the caller of this function is authorized first. A pointer to the loaded key is returned. Only the caller of this function is authorized to use this pointer.
- A method $Id_K \leftarrow decrypt(K^E, Id_{K_{dec}})$, which decrypts the encrypted key K^E with the key referred to by $Id_{K_{dec}}$. The key K_{dec} must be loaded in the TPM's volatile memory beforehand with the $loadKey()$ method. Only a pointer to the key K is returned. The key thus never leaves the TPM. The TPM authorizes the caller of the $decrypt()$ function to execute this function only after having verified the policy corresponding to K^E (see the $storeKey()$ method) and after having verified that the caller of this function is also the owner of the pointer $Id_{K_{dec}}$.
- A method $decryptAndSend(Id_K, content_i^E, Id_{output})$ decrypts the content block $content_i^E$ using the key in the TPM's volatile memory referred to by Id_K and confidentially sends it to the output device with id Id_{output} (after output device authentication). Again, Authorization is given to the caller after having checked the policy file and the ownership of the pointer.
- The method $true/false \leftarrow deleteKey(name)$ removes the key K_{name}, together with the associated data such as its policy, out of the TPM's non-volatile memory. Only the TPM owner or owner of the key (i.e. the one who executed the $storeKey(*, name, *)$ method) can get authorized by the TPM to do this.

An authentication mechanism is required such that the application can authenticate towards the TPM by proving ownership of an embedded key using white box cryptography, but without knowing the key value itself.

We think that these extensions are feasable. Firstly, TCG is planning to extent the TPM specifications with policy support (see [10] section 2.4). Secondly, if we omit the $decryptAndSend()$ method, all the proposed methods are theoretically possible in the most recent TPM specification (V1.2). In fact, besides the policy support, only the API and some access control must be added to the TPM. Thirdly, the most difficult extension seems to be the $decryptAndSend()$ method, because current TPMs do not support mass symmetric encryption. However, hardware implementation of symmetric algorithms such as AES are easy and cheap.

4.4   Protocols

In this section, the different protocols required in the DRM framework are described at a high level. The functionality of the less relevant protocols is only briefly explained. We refer to the full report (see [13]) for all details.

The properties of the communication links are indicated with the letters "I" (integrity), "C" (confidentiality) and "A" (authentication) above the communication arrows. An indirect connection is indicated with an "*". For example, $B \overset{AI}{\leftrightarrow} A$ means that both A and B need to be authenticated, and that the messages, which are sent in both directions, have to be integrity protected. Abstraction is made of the user authentication.

Content submission. The producer negotiates a contract with the Import Service IS before the producer submits the content. IS sends the content to the Content Service CS and the usage rules to the License Service LS.

DRM client installation. To install a DRM client, the Client Setup Service CSS is contacted. CSS embeds a key in the new DRM client using white-box cryptography, applies code guarding and sends the DRM client to the consumer. This key is also imported in the TPM by CSS (hidden for, but with the consent of the consumer) and will be used to authenticate the DRM client to the TPM and to DRM services. The public key of the top level CA (Certification Authority) for DRM services is also imported in the TPM by CSS using the *storeKey* method. The policy states that the key can only be changed by the CSS, but other entities can view it. The DRM client will use it to verify the validity of licenses. The security properties (e.g. code guarding type) of the DRM client can be sent to the Registration Service RS by the CSS.

Device Registration. Devices must be registered before they can consume content. If the DRM system is convinced of the thrustworthiness of the device (in fact the TPM guarding it) and the installed DRM client, the consumer will obtain a per user root license, while the key K_{TPM-C} necessary to extract the root key from the root license is imported in the TPM. The DRM client is only given permission to use this key for cryptographic operations performed inside the TPM. The per user root license key K_{lroot} is encrypted with this key. Only the TPM can thus get hold of the K_{lroot}. RS knows the id of the DRM client and the consumer because these need to be authenticated. By simply registering a device, i.e. by requesting a root license for it, C is thus able to use all his content licenses and content on that device. When a consumer registers his first device, K_{lroot} is first generated (and stored at RS by *retrieveRootKey*.

1	$C \xrightarrow{AI} RS$	$registrationRequest(cert_{TPM})$
2	$TPM \xleftrightarrow{A*} RS$	$proofSystemSecurityStatus()$
3	$TPM \xleftrightarrow{A} client \xleftrightarrow{A} RS$	$checkDRMClientSecurityStatus()$
4	RS	$K_{TPM-C} \leftarrow genKey()$
		$K_{lroot} \leftarrow retrieveRootKey(Id_C)$
		$K_{lroot}^{E} \leftarrow enc_{K_{TPM-C}}(K_{lroot})$
5	$C \leftarrow RS$	$license_{root} \leftarrow send((..., K_{lroot}^{E}, ...)sig_{RS})$
6	$TPM \xleftarrow{ACI*} RS$	$storeKey(K_{TPM-C}, TPM - C, \{Id_{client} : crypto\_use = yes\})$
7	RS	$store(Id_C, Id_{TPM}, date_{currrent})$
8	$LS \leftarrow RS$	$send(Id_C, K_{lroot})$

The $proofSystemSecurityStatus()$ starts a protocol already available in current TPMs. While executing $checkDRMClientSecurityStatus()$, the DRM client proves his identity by proving ownership of the embedded key (using white-box crypto). This is enough for RS to look up the security properties of the DRM client. Optionally, the TPM could be used to further enhance the protocol.

Content request. The consumer C retrieves DRM protected content $content^{WE}$. The protected content is watermarked with the transaction id to enable consumer identification in case of abuse detection. A proof of the transaction details is agreed and stored by CS. Because the proof contains all relevant data, including the transaction id, which are signed by the consumer, this can indeed be considered as a proof. The key to decrypt the content is sent to the License Service LS for inclusion in a license.

$$1 \quad C \xrightarrow{AI} CS \quad request(Id_C, Id_{content})$$

$$2 \quad C \xleftrightarrow{A} CS \quad proof \leftarrow (Id_{content}, Id_C, Id_{transaction}, timestamp)sig_{SK_C}$$

$$3 \qquad CS \qquad content \leftarrow retrieveContent(Id_{content})$$
$$K_{content} \leftarrow genKey()$$
$$content^W \leftarrow watermark_{PK_{IdS}}(content, Id_{transaction})$$
$$content^{WE} \leftarrow enc_{K_{content}}(content^W)$$

$$4 \quad CS \xrightarrow{IA} C \quad send(Id_{content}, content^{WE})$$

$$5 \quad CS \xrightarrow{IAC} LS \quad send(Id_C, Id_{content}, K_{content})$$

License request. Once the consumer has retrieved protected content, he/she can request a corresponding license. After receiving a request, the License Service LS retrieves the rights, encrypts the corresponding content key $K_{content}$ with the per user root license key K_{lroot}. LS indeed has knowledge of both keys. This encrypted key is included in the license. K_{lroot} is generated and sent to RS if it wasn't generated beforehand.

$$1 \quad C \xrightarrow{IA} LS \quad send(Id_C, Id_{content}, Id_{licenseType})$$

$$2 \quad LS \qquad K_{content} \leftarrow retrieveContentKey(Id_{content}, Id_C)$$
$$K_{lroot} \leftarrow retrieveRootKey(Id_C)$$
$$K_{content}^E \leftarrow enc_{K_{lroot}}(K_{content})$$
$$rights \leftarrow retrieveRights(Id_{licenseType})$$

$$3 \quad C \leftarrow LS \quad license \leftarrow (Id_C, ..., K_{content}^E, rights, ...)sig_{LS}$$

Content Consumption. After having obtained a root licence on the device, DRM protected content and a corresponding device license, the consumer will be able to consume the content. The DRM client verifies whether the action is allowed or not. This also includes checking the validity of the licenses. The DRM client retrieves a pointer to K_{TPM-C}, which is used to let the TPM decrypt K_{lroot}^E. K_{lroot} is loaded in the internal volatile memory of the TPM. Only a reference is returned, such that it can be used to decrypt $K_{content}^E$ in a similar way. Once the content key is known by the TPM, *client* sends the content to the TPM which decrypts it and confidentially sends it to the output device.

$$1 \quad C \rightarrow client \quad consume(content^{WE}, license, license_{root}, action, Id_{output})$$

$$2 \qquad client \qquad actionAllowed(license, license_{root}, action, Id_{output})$$

$$3 \quad client \xrightarrow{AC} TPM \quad Id_{K_{TPM-C}} \leftarrow loadKey(TPM - C)$$
$$Id_{K_{lroot}} \leftarrow decrypt(Id_{K_{TPM-C}}, license_{root}.K_{lroot}^E)$$
$$Id_{K_{content}} \leftarrow decrypt(Id_{K_{lroot}}, license.K_{content}^E)$$
$$decryptAndSend(Id_{K_{content}}, content^{WE}, Id_{output})$$

Identification. The identification Service IdS is the only entity that is able to identify the source of abuse when unencrypted but watermarked content is found by extracting the watermark. The embedded transactioon id can be used to retrieve the transaction details signed by the consumer.

Device Deregistration. Old devices can be unregistered to prevent further consumption of DRM content on that device. The DRM client sends a request to the Registration Service RS. RS establishes a confidential connection with the TPM and removes K_{TPM-C} (and associated data such as the policy) stored by the TPM by executing the $deleteKey(TPM - C)$ method on the TPM. Once this is done, RS locally removes the registration tuple.

Rights often are time related. The DRM client thus needs a tamper resistant clock. The DRM client is able to detect tampering with the system clock by using a tick counter and a monotonic counter, complemented with an online timing service (see the report [13] for details).

5 Analysis

Leaking secret key data. If no extra protection mechanism is present, which is the case in current computers, the content of the internal memory can be leaked by doing memory dumps, or by reading swap data. At the consumer's side, the sensitive key information is only in clear text in the shielded volatile memory of the TPM. Only the DRM client is authorized to request decryptions with the key, without ever having access to the key itself. The online DRM system also knows secret key information (e.g. K_{lroot}). Classical protection mechanisms are required here. The confidentiality of the symmetric key used by the DRM client to authenticate, depends on the robustness of the applied white-box cryptography algorithm. Keys are always transferred confidentally. It must not be possible to extract DRM keys out of the TPM.

Leaking unencrypted content. Measures must be taken by the publisher and producer to keep the content secret using classical cryptography. The only place at the consumer side where content (or the decrypted keys) resides unencrypted is in the TPM and in the output device. We assumed that the latter is sufficiently protected. Thus, even if the DRM client is broken, the consumer or an attacker cannot get hold of the content.

Spreading of recorded content. It is impossible to avoid recording of content once it leaves the output device (e.g. recording audio). This is called the analog gap, which is bridged by inserting a watermark that contains the transaction identifier. The spreading of content relies on the robustness of the watermarking scheme that is used.

Rights extension or theft. We rely on existing code guarding techniques to detect tampering with the functionality of the DRM client. If the code guarding is broken, the rights can no longer be enforced. This may allow the consumer to extend his 'rights' in an illegal way. Still, the consumer can not get hold of the content or sensitive key information. If a set of DRM clients is considered insecure, the consumers can be forced to update their DRM client when a new root license is required. Typically, the lifetime a root license will be rather short. This is not necessarily an extra burden to the consumer, because it can be made invisible for him/her.

During the client installation, the TPM is requested to store the public key of the DRM public key of the top level CA. The certification chain verification of a root license or content license does not succeed without this public key. Thus, the consumer cannot take the encrypted keys out of valid licenses and put them into e.g. self signed licenses with more rights.

Only C can obtain a root license for his/her devices. If no consumer authentication is required before content can be consumed, C can simply give another consumer C' illegal access to content by placing a root license on C''s device. Therefore, it is important that the number of root licenses that C can have at the same time is limited and that C has to authenticate to the DRM client on (semi) public computers before he/she can consume content.

The clock tampering detection mechanism avoids that consumers consume content after expiry of the content license or root license. When a root license is found expired, the Registration Service RS will request the TPM to remove the associated symmetric key. The registration record stored by the Registration Service will also be deleted after expiry.

If a DRM client is found compromised or insecure, RS will try to establish a connection with the TPM in order to remove K_{TPM-C} the next time the device connects to the internet. A new root license will only be issued after having updated the DRM client. However, this cannot always be enforced and consent of the TPM owner will be required.

With the current technologies, the services can be convinced of the trustworthiness of the system and the DRM client. The DRM keys can be stored by the TPM. However, cryptographic functions are performed by software. Keys and output of these cryptographic functions are stored in the internal memory, which can be obtained by the consumer. The DRM extensions presented in this paper thus offer a considerable increase towards secure DRM. In current solutions, the security is often based on code obfuscation. Knowing the hidden algorithms enables consumers to extract the content if a valid license is present. This framework does not rely on secrecy of algorithms, but only on secrecy of keys.

If a trustworthy DRM client is replaced by a non trustworthy one, the latter will not succeed in getting authorization by the TPM when trying to consume content, because it lacks the white-box crypto embodiment of the key needed to authenticate..

6   Comparison with Existing Technologies

More and more, DRM is apprearing in new products such as Compact Disks and MP3 players. All companies owning the technologies try to hide as much details about the inner working as possible. Recently, we have seen Sony's DRM technology on CDs being critisized for creating hidden files on the consumer's system and for running secretly processes that can compromise the system's stability. These files are hard to remove without losing access to your CD-drive. If we look at Microsoft's Windows Media DRM and Apple's Fairplay (used in iTunes and iPod), we see the same: they do not offer much information about the inner working and trust on the secret keeping of algorithms that are used by their DRM clients. E.g. Windows Media DRM uses code obfuscation to hide the algorithm that derives the DRM client key. Fairplay has similar problems. Sooner or later, these algorithms will be discovered, resulting in the DRM technologies being broken.

One important aspect in this paper is that we provide an open DRM system; anyone may know the inner working. We can compare the evolution in DRM with the evolution of cryptography in general where we saw a change from hiding algorithms to hiding keys. Our framework uses several building blocks that can be replaced if necessary. At the moment, only Windows Media DRM offers limited updates for its DRM clients, however if the algorithms are revealed, more than a simple update is required. Another important aspect of this framework is the binding of rights to users instead of to devices.

The presented framework needs more hardware support, which is indispensable to have secure DRM. The key information must not be exposed, which is only possible with hardware or operating system support. Microsoft is working on NGSCB (see [9]), which should offer operating system support similar as the operating system based solution presented in [11]. In the case of DRM, the consumer must be seen as a potential adversary. Therefore, we think that it is hard to combine open source and DRM support in operating systems. The presented solutions indeed requires extra hardware support, but can still be used on open source operating systems such as Linux.

7   Conclusion and Future Work

In this paper, we presented a flexible DRM framework that is more secure than existing technologies. To have a secure DRM technology that cannot be broken by consumers or other attackers, we need protection against memory space snooping. This can be done by hardware or operating system support. This paper presented a hardware based solutions, whereof we think that these allow a greater degree of openness and simplicity than the operating system approach. The hardware based solution needs some extensions to the current TPM (v1.2). We argued that the current TPM offers a good basis such that minimal extensions satisfy to allow secure DRM systems. This paper can be seen as a proposal towards TCG to extend the TPM for DRM purposes. Of course, appropriate watermarking, code guarding and white box cryptography is required. These technologies can only be seen as blocks in a complete DRM system. Only the most essential key information is hidden while at the same time the functionality of the DRM client is protected. This paper tried to identify the crucial protocols. Extra services and protocols can be added. The paper is high level, but hopefully, it will help in developing mature DRM.

As part of future work, stateful licenses will be taken into account. These limit the number of times certain actions on the content can be performed by the owner of the license. Conditional anonymity will be added, such that the identity of both producer and consumer is not revealed in the case of normal usage. This will be applied in the domain of e-health. Delegation of rights and DRM client revocation will also be tackled. This should allow the owner of rights to lend, give or sell part of these rights to others.

References

1. M. J. Atallah, E. D. Bryant, M. R. Stytz. A survey of Anti-Tamper Technologies. 2004.
2. Mauro Barni and Ingemar J. Cox and Ton Kalker and Hyoung Joong Kim. Digital Watermarking, 4th International Workshop, IWDW 2005, Siena, Italy, September 15-17, 2005, Proceedings.
3. S. Chow, P. Eisen, H. Johnson and P.C. van Oorschot. A White-Box DES Implementation for DRM applications. In Proceedings of ACM CCS-9 Workshop DRM 2002, volume 2595 of Lecture Notes in Computer Science, pages 1-15. Springer-Verlag, 2003.
4. S. Chow, P. Eisen, H. Johnson and P.C. van Oorschot. White-Box Cryptography and an AES implementation. SAC 2002 - 9th Annual Workshop on Selected Areas in Cryptography, Aug.15-16 2002, St. John's, Canada. Proceedings (revised papers): pp.250-270, Springer LNCS 2595, 2003.
5. Bill Horne, Lesley Matheson, Casey Sheehan, and Robert E. Tarjan. Dynamic Self Checking Techniques for Improved Tamper Resistance. ACM Workshop on Security and Privacy in Digital Rights Management, pages: 141 - 159, 2001.
6. William Ku and Chi-Hung Chi. Survey on the Technological aspects of Digital Rights Management. ISC 2004: 391-403, 2004.
7. Windows Media DRM 10 cracked?. http://www.engadget.com/2005/02/01/windows-media-drm-10-cracked/, 2005.
8. S. Michiels, K. Verslype, W. Joosen, B. De Decker. Towards a software architecture for DRM. DRM '05: Proceedings of the 5th ACM workshop on Digital rights management, 2005.
9. Microsoft Next Generation Secure Computing Base http://www.microsoft.com/resources/ngscb/default.mspx.
10. G.J. Proudler. Concepts of Trusted Computing. IEE Professional Applications of Computing Series 6, 2005.
11. J. F. Reid, W. J. Caelli. DRM, Trusted Computing and Operating System Architecture. 2005.
12. Trusted Computing Group. TCG TPM Specification Version 1.2 Revision 94, https://www.trustedcomputinggroup.org/specs/TPM, 2006.
13. K. Verslype, B. De Decker. A Flexible and Open DRM construction. KULeuven dept. computer science, technical report, 2006.

PPINA - A Forensic Investigation Protocol for Privacy Enhancing Technologies

Giannakis Antoniou[1], Campbell Wilson[1], and Dimitris Geneiatakis[2]

[1] Faculty of Information Technology, Monash University, Caulfield East Melbourne,
3145 Victoria, Australia
gant2@student.monash.edu.au, cambell.wilson@infotech.monash.edu.au
[2] Dept. of Information and Communication Systems Engineering, University of the
Aegean, Karlovassi, Samos, 83200 Greece
dgen@aegean.gr

Abstract. Although privacy is often seen as an essential right for inter-net users, the provision of anonymity can also provide the ultimate cover for malicious users. Privacy Enhancing Technologies (PETs) should not only hide the identity of legitimate users but also provide means by which evidence of malicious activity can be gathered. This paper proposes a forensic investigation technique, which can be embedded in the frame-work of existing PETs , thereby adding network forensic functionality to the PET. This approach introduces a new dimension to the implementa-tion of Privacy Enhancing Technologies, which enhances their viability in the global network environment.

Keywords: Network Forensics, Privacy Enhancing Technologies.

1 Introduction

Privacy Enhancing Technologies (PETs) provide an environment for internet users within which connection anonymity [1] can be assured. However, the pro-vision of such anonymity without proper control has the potential to cause chaos within the internet society rather than helping legitimate users. Generally, in-ternet users want to be able to take advantage of the network services offered by the internet without having to necessarily reveal their identity. On the other hand, servers providing such services should also have a mechanism by which the identity of any malicious user (for example a user taking part in an attack on network resources) can be unveiled if necessary, and evidence of such a user's activity provided to the appropriate entity.

The field of network forensics involves the investigation of cyber-crimes, in-cluding establishing the identity of internet abusers and gathering evidence of malicious activity for presentation to law courts. A key component in network forensics, which provides strong evidence of identity, is the digital signature. In offline life, the written signature is an important aspect of an agreement between

H. Leitold and E. Markatos (Eds.): CMS 2006, LNCS 4237, pp. 185–195, 2006.
© IFIP International Federation for Information Processing 2006

two people. In the digital society, the digital signature is an equivalent way of legally enforcing an agreement between two parties [15]. In the United States [2,3,4] and the European Union [4,5], written signatures and digital signatures have the same legal standing. The techniques we propose in this paper involve the use of the digital signature as a means of identifying users and increasing the level of non-repudiation bestowed on the actions of a user.

Research efforts into PETs and network forensics have essentially opposite respective goals. While PETs attempt to hide the identity of users, network forensics is an area primarily concerned with the revealing of the identity of abusers. The philosophy behind the PPINA (Protect Private Information, Not Abuser) technique presented in this paper involves bridging these two research areas in order to produce a harmonious combination, which can serve both legitimate internet users as well as law enforcement agencies. We have to make it clear at this point that the purpose of the paper is not to introduce a new PET, but to introduce a forensic investigation technique, which can be embedded in a PET framework.

The underlying scenario for our technique is that any user can be anonymous (i.e. protected by the PET) unless the Server requests a forensic investigation entity (FIE) to investigate a particular sequence of packets received by the Anonymous User (AU) through a PET. If the server has enough evidence to prove that somebody has tried to attack the server, then the FIE will further investigate and reveal the identity of the abuser. At the end of this process, a strong body of evidence will be built up concerning the abuser's (non-repudiated) actions. To the authors' knowledge, this is the first time network forensics and privacy enhancing technology have been combined in a single unified framework.

The paper is organized as follows: Section 2 examines the general framework of PETs; Section 3 outlines the motivations of the proposed solution; Section 4 introduces our proposed technique; Section 5 presents a hypothetical case study illustrating the application of our proposed technique and Section 6 concludes the paper.

2   The General PET Framework

In the last 2 decades, several PET protocols [6,7,8,9,10,11,12,13] offering anonymity, have been proposed. Most of them do successfully offer privacy (at the network layer) by hiding the users' IP addresses. However, none of them includes techniques for revealing the identity of those users who are abusing the network resources and gathering supporting evidence of such activities.

Although there is a plethora of PETs, most of them have the same frame-work (figure 1). Each PET protocol is distinguished based on the algorithm used to forward anonymously the messages from the ClientA (Anonymous User) to the ServerB (Server) and back again.

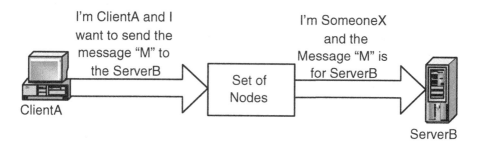

Fig. 1. General PET Framework

3   Motivations

There is no current technology that offers anonymity to a user and at the same time discourages that user to act maliciously against the server. In particular, none of the related PET technologies offer:

a) **Complete non-repudiation:** In order for a Server to be able to accuse an abuser, the server needs strong evidence about the action of the abuser. Without such evidence, the abuser cannot be prosecuted. The ultimate evidence in the digital world is the digital signature [15] because it assures an action cannot be repudiated by the abuser.

b) **Complete confidentiality and integrity:** An internet user wants to be anonymous and only the destination server must read his or her messages. No one else, even the trusted Privacy Enhancing Entities, must be able to access them, i.e. the integrity of the messages must be protected. However, at present, PETs have at least one node that has access to the unencrypted messages and no technique has therefore been employed to offer complete integrity of the messages.

c) **Complete authentication:** Internet users have recently seen anonymity as an important facet of network communication [14]. At the same time, for many years, malicious intruders have also been looking for such anonymity. Intrusion is an illegal action and an intruder therefore wants to become anonymous (for different reason than an internet user) during his or her activities. Therefore, an intruder can use the PET to hide his or her identity, and the PET helps (unknowingly) the intruder. For this reason, the PET should be sure that the client is the person that it claims to be, before offering anonymity to the client. Although the PET may offer anonymity to the client without the necessity for identifying the client, the PET can become a very good tool for any intruder. Of course, an intruder can also hide his identity illegally, without using a PET framework (for example, by spoofing his or her IP address). However, the PET legitimates in a sense this identity hiding action. For this reason, a mechanism should be applied to identify only the abusers.

These above issues are essential considerations in the design of an appropriate framework, which offers network forensics services in a PET framework. The next section presents the proposed framework design and the communication protocol, which add network forensics services in the PET framework.

4   The *PPINA* FRAMEWORK

The PPINA (Protect Private Information Not Abuser) framework offers a proactive forensic investigation technique that can be embedded in any PET because it is independent of a PET protocol. It adds an end-to-end confidentiality and integrity layer (Issue (b) from our motivations) and forensic investigation service (Issues (a) and (c) from our motivations). In addition, the Server does not have its functionality compromised during the forensic investigation. The Server only needs to contact the Forensic Investigation Entity (FIE) and send the malicious messages. The FIE verifies the authenticity and the integrity of the messages as well as whether the messages are malicious or not. In case the FIE concludes that the messages are malicious, it replies with evidence (which proves the involvement of the attacker and cannot be repudiated) and the identity of the attacker. We emphasize again that the PPINA protocol operates over a PET protocol and is therefore a general solution, not linked with a specific PET. The following explains the operation of the PPINA protocol. We first introduce the notation used in the explanation.

Notation

A = Anonymous User
B = Directory Service
C = PET
D = Server
$As\{Data\}$ = The Data is signed by the private key of an Anonymous User, where the public key, of that private key, is published
$Ae\{Data\}$ = The Data is encrypted by the public key of an Anonymous User, where the public key is published
$s\{Data\}$ = The Data is signed by the private key, which is created for the needs of a session. The public key of that private key is not published. Only the Server and the AU know that public key. This public key plays, also, the role of a secret key
$e\{Data\}$ = The Data is encrypted by a secret key (symmetric encryption)
$bKey\{Data\}$ = The Data is encrypted by the bKey (symmetric encryption)
$ForensicReceipt$ = The digest of the received data from the Server

4.1   The Three Phases

The whole communication process can be divided into 3 phases: the Initialization phase (Figure-2, Steps 1-6), the Main phase (Figure-2, Steps 7-10) and the Forensic Investigation phase (Figure-3, Steps 11-15).

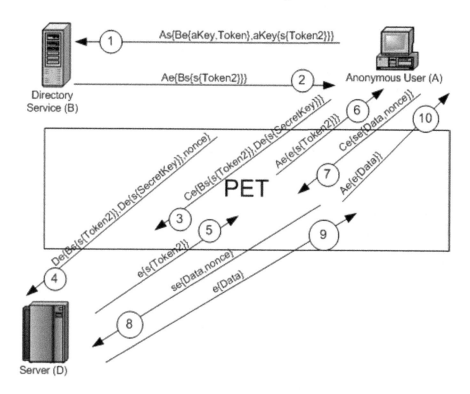

Fig. 2. PPINA (Initialization and Main Phase)

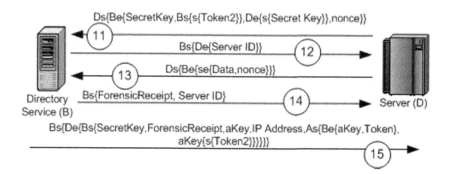

Fig. 3. PPINA (Forensic Investigation Phase)

Initialization Phase

Before the real communication (main phase) begins, the initialization phase is required. The DS gathers the fingerprint [s{Token2}] of the future communication actions of the AU, and the Server validates that fingerprint:

A→B: As{ Be{ aKey, Token}, aKey{ s{ Token2 } } } **(Step 1)**
B→A: Ae{ Bs{ s{ Token2} } } **(Step 2)**
A→C: Ce{ Bs{ s{ Token2} }, De{ s{ Secret Key} } } **(Step 3)**
C→D: De{ Bs{ s{ Token2} }, De{ s{ Secret Key} }, nonce} **(Step 4)**
D→C: e{ s{ Token2} } **(Step 5)**
C→A: Ae{ e{ s{ Token2} } } **(Step 6)**

The AU generates a pair of keys (Public/Private), where the public key plays also the role of a secret key (for symmetric encryption) in order to provide end-to-end data encryption [e{Data}]. This secret key is valid only during that session. After the end of that session, the secret key is invalid, and a new pair of keys should be generated for future sessions, even if the participating entities are the same. The AU signs with the private key and encrypts with the secret key [se{Data}]. The Server can verify and decrypt the data [se{Data}] with the secret key.

The AU calculates the Token [Token= Hash_Function(Secret Key)] and the Token2 [Token2= Hash_Function(Token)]. The AU signs the Token2 [s{Token2}] by using the private key. In addition, the AU generates a symmetric key (aKey) to encrypt part of the data sent to the DS. The [s{Token2}] is encrypted (Step 1) in order to avoid a possible attack from the Server. The DS is responsible for verifying the validity of Token2, based on the given Token. The DS stores the message (Step 1) in order to prove, in case of a forensic investigation, that the specific AU was going to communicate with a Server by using a secret key, which has the specific Token2. The [Bs{s{Token2}}] is the ticket which is forwarded to the Server through the AU and the PET. The AU also sends [De{s{Secret Key}}] to the Server. The Server is responsible to verify the validity and authenticity of Token2, based on the given secret key. The Server stores the message (Step 4) in order to prove later on (to the Forensic Entity) that the communication has used a secret key with a specific Token2. Successful verification of the DS by the Server means that the secret key is linked with the Token2 (Secret Key → Token → Token2), which Token2 is linked with the AU who has signed the message (Step 1). However, the Server needs to verify that the AU also has the private key by verifying the signature of the secret key [s{Secret Key}] and [s{Token2}] by using the secret key as a public key.

Generating private/public key pairs is a computationally intensive procedure. However, an AU can generate several pairs of keys during times of low system load and request tickets from the DS (one ticket for each pair of key). A ticket is server-independent and time-independent; therefore, a ticket can be used for future communication with any Server and any time. A ticket is useful only to the entity who knows the related private key.

Main Phase
Once the AU and the Server have exchanged the secret key and the Server has validated and authenticated the value of the Token2, the session has been established and the Main Phase is ready to begin. The AU can enjoy the

confidentiality/integrity of the exchanged messages and the anonymity offered by the PET, while the Server is ensured that the identity of the AU will be revealed (from the FIE) in case the AU is an abuser:

A→C: Ce{ se{ Data, nonce} } (**Step 7**)
C→D : se{ Data, nonce} (**Step 8**)
D→C : e{ Data} (**Step 9**)
C→A: Ae{ e{ Data} } (**Step 10**)

The AU encrypts, with the public key (secret key), and signs, with the private key, the message [Data, nonce]. Each message (Step 7) contains the Data and the nonce. The nonce is a counter, which helps to avoid a replay attack. If the Server receives an encrypted message (data and nonce) twice, the Server rejects the last message. The nonce is also useful during the Forensic Investigation Phase, when the FIE tries to determine whether a series of packets are malicious or not.

The Server stores the message (Step 8) in order to prove, in case of an attack, that the AU who has the private key of the public key (secret key) has sent the message. The Server decrypts and verifies the message [se{ Data, nonce}] by using the secret key. In case that the Server wants to reply, it will encrypt the data with the secret key, and it will send the data to the AU (Step 10) through the PET (Step 9).

Forensic Investigation Phase
During this phase, the DS plays the role of the Forensic Investigation Entity (FIE). In case the Server receives inappropriate data from an abuser, the Server informs the appropriate entity (FIE), which will investigate the incident and identify the abuser. The FIE needs evidence of the abuser's action in order to continue the investigation. The Server must provide such evidence. The Server should have saved the communication messages between the PET and Server to prove that the messages came from the specific PET. The Server accuses the PET, until the PET provides evidence that an Anonymous User generated these messages. The FIE should provide evidence about the identity of the Anonymous User.

It is particularly important that in our framework, the Forensic Investigation Phase can take place while the Server is still functioning. In current network forensic investigations, it is usual for the Server needs to stop functioning for days after an incident, while the forensic entity investigates the Server for evidence.

During the Forensics Investigation phase, none of the entities needs to reveal their private keys in order to prove their claims. However, the Server is required to reveal the exchanged secret key used during the communication with the malicious user.

D→B: Ds{ Be{ Secret Key, Bs{ s{ Token2} }, De{ s{ Secret Key} }, nonce} }
(**Step 11**)
B→D: Bs{ De{ Server ID} } (**Step 12**)

D→B: Ds{ Be{ se{ Data, nonce} } } **(Step 13)**
B→D: Bs{ ForensicReceipt, Server ID} **(Step 14)**
B→D: Bs{ De{ Bs{ Secret Key, ForensicReceipt, aKey, IP Address, As{ Be{ aKey, Token}, aKey{ s{ Token2} } } } } } **(Step 15)**

The Server sends (Step 11) the ticket (generated and signed by the DS) to the FIE including the secret key. The FIE verifies the validity of the secret key (Secret Key → Token) and the authenticity of the Token2 [s{Token2}] and replies (Step 12) with a Server ID. This identity (Server ID) is the identification of the current investigation case. The Server sends to the FIE (Step 13) all the communication messages for that session received by the AU. After the FIE receives all the necessary messages, it replies (Step 14) with a ForensicReceipt. The ForensicReceipt is the hash value of the (Step 11) and (Step 13). The Server, also, calculates the ForensicReceipt. These two ForensicReceipts should have the same value. Otherwise, there is a problem with the integrity of the exchanged messages. The ForensicReceipt ensures that the FIE received all the data that the Server has sent. The FIE, also, cannot deny that the Server asked the FIE to investigate the case (The FIE signs the ForensicReceipt).

After the FIE concludes that the messages were malicious, the FIE sends (Step 15) the IP Address of the AU and evidence (Step 1) which proves the involvement of the AU. Only an entity who knows the private key could sign the Token2. The Server can now submit [Bs{ Secret Key, ForensicReceipt, aKey, IP Address, As{ Be{ aKey, Token}, aKey{ s{ Token2 } } } }] from (Step 15), and the malicious communicated messages from the AU to the server (Step 13) to the court as evidence of the AU's actions.

We have described the operation of the PPINA protocol. The PPINA technique embeds in a PET framework a number of characteristics, which are described in detail in the next section.

4.2 Characteristics of the PPINA Protocol

Provision of Strong Evidence: Since the messages have been signed by the AU, it can be confirmed the actions of the AU/abuser cannot be repudiated by the AU. The digital signature provides the ultimate legal means of evidence verification in the digital era; therefore, no entity is able to doubt about the integrity and the authenticity of the evidence.

Non-stop Server/victim operation: In a classic scenario, during an investigation, the Server/victim needs to be investigated closely by experts in order to gather evidence about the actions of the abuser. During the investigation, the Server/victim is typically not in online fully operational mode. However, in this case, the Server's computer does not need to be investigated, because the DS has the necessary evidence to accuse the abuser.

Related cheap and fast investigation procedure: The PPINA protocol forces every AU to provide evidence to the DS of the AU's future actions before contacting the Server. Therefore, the Server knows that the DS has the necessary evidence, making the investigation not only fast but also cheap.

No privacy violation exists during the investigation: As part of the computer forensic investigation procedure, the Server does not need to make available any storage media (Hard Disk, Tapes, CD-Rom, etc) which took place during the cyber-crime, which may also contains private sensitive information of the Server. **Respect the goal of the underlying PET:** The PPINA can be embedded in a PET framework without to affect the level of offered privacy of the PET.

Our proposed framework offers significant advantages. However, it should be acknowledged that the level of encryption involved and the imposition of the authentication layer might decrease the level of performance of the communication between AU and Server. In addition, if the Server somehow compromises the DS, the Server can identify the users who have contacted with that Server.

5   Case Study - AU Attacking a Server

In this section, we explain in more detail the specific operation of the PPINA protocol using a simple case study whereby an anonymous user attempts to attack a Server via the PET. Suppose Alice (Anonymous User) wants to communicate with Bob (Server). Firstly, she generates a pair of keys (Public/Private) and then contacts the DS in order to get the necessary ticket [Bs{s{Token2}}], which is mandatory for the communication between Alice and Bob. Before the DS issues the ticket, it verifies that the Token2 is the hash value of the Token. Otherwise, the DS does not issue the ticket. Alice, through the PET, forwards the ticket to Bob. Alice, also, creates [De{s{Secret Key}}] and sends it to Bob. The Secret key (public key) is needed to offer confidentiality of the messages and also to verify the signed messages of Alice. Bob verifies that the Token2 is the hash value of a token, where token is the hash value of the secret key. Also Bob verifies the signature of [s{Token2}] and [s{Secret Key}] with the Secret Key, which here plays the role of the Public Key. If the verifications are valid, Bob encrypts the [s{Token2}] and sends it back to Alice, otherwise Bob terminates the communication. After the Initialization Phase is completed, the Main Phase begins whereby Alice wants to compromise Bob's computer. Alice signs the message with the private key and encrypts it(via symmetric encryption) with the secret key. Bob receives the message and makes the necessary verification, whereby he decrypts the message and verifies the signature with the secret key. Once Bob realizes the attack, he stops the communication with Alice, contacts the DS and sends the appropriate evidence (all the information received by Alice including the secret key). The DS cannot deny the existence of Alice because the DS has issued the ticket (there is a signature of the DS on the ticket). The DS cannot also accuse an innocent AU (i.e. another anonymous user other then Alice) because Bob has the ability to verify the [s{Token2}] via the secret key. Bob expects to receive, from the DS, a signed message that includes a specific [s{Token2}], which can be verified by the secret key. The DS decrypts and verifies the messages with the secret key and examines the information [se{Data,nonce}]. If the DS detects that the information was malicious, it replies with the evidence [As{Be{aKey, Token}, aKey{s{Token2}}}] and additionally sends the IP Address of the user as well as the aKey. aKey will be used to decrypt the encrypted message in order Bob verifies the signature [s{Token2}] and is therefore sure about the identity of the abuser. It is possible that a compromised DS can send wrong IP Address. However, the user who has signed the message [As{Be{aKey,Token},aKey{s{Token2}}}] is the abuser, because only this AU knows the Private Key (AU signed Token2). The Server now has the strongest evidence to prove the involvement of that particular user in the attack.

6   Conclusion

The provision of privacy and anonymity to internet users can also provide an environment within which malicious users can hide. A key driver of the wider adoption of privacy enhancing technologies would be the ability of forensic entities to gather evidence of malicious activity while legitimate users are still offered anonymity. In this paper, we have provided a framework through which the anonymity of users not engaged in malicious activity is protected while such evidence can be gathered when network abuses occur. The PPINA framework offers non-reputation actions of the users (by using the digital signature), adds a layer of message confidentiality (by using a secret key), respects the privacy information of the Server during the investigation, and decreases the cost and the duration of an investigation.

References

1. Andreas   Pfitzmann   (2005),   "Anonymity,   Unlinkability,   Unobservability, Pseudonymity and Identity Management - A Consolidated Proposal for Terminology"
2. http://slis.cua.edu/ihy/fall01/tpedoc/pl106229.pdf
3. http://www.ntia.doc.gov/ntiahome/ntiageneral/esign/105b/esign7.htm
4. Stephen E. Blythe, Digital Signature Law of the United Nations, European Union, United Kingdom and United Stats: Promotion of Growth in E-Commerce with Enhanced Security, 11 RICH. J.L. & TECH. 2 (2005), at http://law.richmond.edu/jolt/v11i2/article6.pdf
5. http://europa.eu.int/eur-lex/pri/en/oj/dat/2000/l_013/l_01320000119en00120020. pdf
6. Gritzalis S., "Enhancing Web Privacy and Anonymity in the Digital Era", Information Management and Computer Security, Vol.12, No.3, pp.255-288, 2004, Emerald
7. Anonymizer (2003), available at http://www.anonymizer.com
8. Reiter M., Rubin A., "Crowds: Anonymity for web transactions", ACM Transactions on Information and System Security (TISSEC), Vol. 1 , Issue 1 (Nov 1998), Pages: 66 - 92
9. Roger Dingledine, Nick Mathewson, and Paul Syverson. Tor: The Second-Generation Onion Router. In Proceedings of the 13th USENIX Security, Symposium, August 2004

10. Clay Shields, Brian Neil Levine, "A Protocol for Anonymous Communication Over the Internet", November 2000 Proceedings of the 7th ACM conference on Computer and communications security
11. Philippe Golle, Ari Juels, "Parallel Mixing", October 2004 Proceedings of the 11th ACM conference on Computer and communications security
12. Ulf Moller, Lance Cottrell, Peter Palfrader, and Len Sassaman. "Mixmaster Protocol", Version 2. Draft, July 2003, available at http://www.abditum.com/mixmaster-spec.txt
13. Marc Rennhard, Bernhard Plattner, "Practical anonymity for the masses with morphmix", In Ari Juels, editor, Financial Cryptography. Springer-Verlag, LNCS 3110, 2004.
14. Alessandro Acquisti (2005), "Privacy in Electronic Commerce and the Economics of Immediate Gratification"
15. Patrick W. Brown , "Digital signatures: can they be accepted as legal signatures in EDI?", December 1993, Proceedings of the 1st ACM conference on Computer and communications security

A Privacy Agent in Context-Aware Ubiquitous Computing Environments

Ni (Jenny) Zhang and Chris Todd

Department of Electronic & Electrical Engineering,
University College London
London WC1E 7JE, United Kingdom
{jenny.zhang, c.todd}@ee.ucl.ac.uk

Abstract. This paper targets personal privacy protection in context-aware ubiquitous computing environments. It proposes a privacy agent technology to help notify people of relevant information disclosure, and to empower them to manage privacy with relative ease. In essence, the development of the privacy agent technology employs privacy terminology and policies specified in Platform for Privacy Preferences Project (P3P) [1], and uses ontological modeling technique to facilitate automated processes of privacy-relevant interactions on behalf of individuals. The development of privacy agent is an integrated part of our ongoing effort towards developing a privacy-respecting context-aware infrastructure.

Keywords: Privacy Protection, Context-Awareness, Ubiquitous Computing, Ontology, P3P.

1 Introduction

In ubiquitous computing environments, sensors and embedded computing devices make it easier than ever to collect and use information about individuals without their knowledge. This has led to a great privacy concern about the potential for abusing personal sensitive information, unease over a potential lack of privacy control, and general desire for privacy-respecting systems [2]. Privacy problems only worsen in context-aware paradigm, where the ubiquitous computing environments discover and take advantage of contextual information (such as user activity, location, time of day, nearby devices) to make decisions about how to dynamically provide services to meet user requirements. Under this circumstance, information that can be used to characterize privacy aspects of an individual comes from various types of sources and with different sensitivity. It is likely that individual privacy preferences towards the dynamic context-aware environment comprise a complex set of rules in response to various situations and changes over time. These make it challenging to provide an adequate privacy protection therein.

Unfortunately, existing approaches focusing on conventional data management environments are inadequate to support dynamic privacy requirements presented in context-aware paradigm. Most of the privacy efforts in the field of ubiquitous computing have been concerned with integrating access control mechanisms into ubiquitous computing infrastructure [3,4,5,6], and employing conventional encryption

H. Leitold and E. Markatos (Eds.): CMS 2006, LNCS 4237, pp. 196–205, 2006.

and security mechanisms as well as identity management tools (such as anonymity and pseudonymity techniques) to complete privacy protection [3,7]. These solutions addressed parts of privacy challenges faced in context-aware systems, but did not support active participation and choice of individuals to control over their personal data. Quite often people are allowed to specify their privacy requirements only by filling in some forms with predefined layout and options, as exemplified in [4, 8]. Such fairly simple approach would be not useful where a person's willingness to share personal information may depend in part on time, his location, and current activities, and may change over time. Demands for flexible mechanisms and user interface for relatively unobtrusive user participation in controlling information disclosure (including getting notice, feedback, and explicit consent) are significant.

In this paper, we propose an intelligent agent to handle privacy-related interactions on behalf of individuals. The privacy agent aims at addressing two key concerns of preserving privacy in context-aware ubiquitous computing environments: privacy feedback (notifying people of relevant information disclosure) and privacy management (i.e. allowing people to express their privacy preferences and manage privacy levels). The development of the intelligent privacy agent is characterized by developing automated preference mechanisms, considering that the task to take full context-aware controls over how personal information is shared can be overwhelming to individuals and might disrupt their ongoing activities, which could defeat the basis goal to make context-aware environments unobtrusive.

The paper is structured as follows. Section 2 presents a use scenario showing how people could use the envisioned privacy agent to preserve their privacy. In section 3, we introduce briefly a privacy-respecting context-aware architecture prototype of which the design of the privacy agent is an integrated part. Section 4 presents a privacy vocabulary and describes how we use ontological modeling techniques to model the privacy vocabulary, in order to facilitate automated processes of the privacy agent. Section 5 continues with some implementation consideration of the privacy agent. In section 6, we look at relevant research efforts towards privacy protection in ubiquitous computing environments. The paper ends with section 7 where a summary of this paper and a brief description of future work are presented.

2   Use Scenario of Privacy Agent

Imagining a wireless-networked city offers context-aware ubiquitous computing services. The city's tourist information center provides a location-tracking service so that tourists can use personalized shopping-guide applications in each shop.

Alice is a tourist visiting the city and carries her smart phone in order to use context-aware ubiquitous computing services. The smart phone serves as a personal assistant and provides Alice an interface to specify her privacy preferences. The privacy preferences are uploaded to and stored at Alice's Privacy Agent (PA) residing somewhere on network.

It is assumed that Alice has specified that any services or applications can use pseudonyms stored on her smart phone to deliver personalized services without alerting her, while any services or applications requiring her real identity and exact location must have her explicit consent.

As soon as Alice enters the city's tourist information center, the location-tracking service advertises itself. The advertisement states tourist guide applications that Alice will benefit from, as well as accompanying data collecting policies which specify data collectors, requested information with desired level of granularity, intended use, expected duration of use, potential third parties, etc. Alice's PA reads the collecting policies, compares them with Alice's privacy preferences. A conflicting interest is detected as the location-tracking service asks for Alice's exact location in order for tourist guide applications to function. The PA then notifies Alice (through her smart phone) of the privacy conflict and wait for her approval or rejection. In case no conflict of interest is detected, the Privacy Agent will not intrusively notify or alert Alice.

Alice then finds that the service offer is interesting and replies to her PA that she would like to accept the service offer in compromising her wish for privacy. Then, when Alice walks into a supermarket, a personalized advertisement service based on Alice's personal profile (e.g. gender, age, occupation, purchase history, etc.) is offered. Alice's PA recognizes the need of a unique identity to use this service, but continues to respect Alice's privacy by offering a pseudonym in place of her real identity. Only when Alice checks out, the Privacy Agent uses Alice's credit card (with real identity information) for payment.

3   A Privacy-Respecting Context-Aware Architecture Prototype

The above scenario outlined basic notions of preserving privacy in context-aware ubiquitous computing environments. To work with the scenario, we have developed a privacy-respecting context-aware architecture prototype within which privacy agents play an important role in implementing privacy protection mechanisms.

As illustrated in Figure 1, a layered architecture and components framed by broken lines in the right of the figure present an architectural support for developing context-aware systems. It provisions four key functionalities of Context Collecting, Storage,

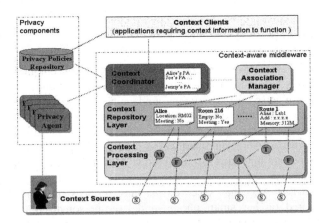

Fig. 1. Context-awareness architecture prototype and privacy components

Processing and Dissemination [9]. Context Processing Layer is responsible for manipulating raw context data to appropriate levels of abstraction that are desired by context clients; Context Repository Layer provides ability to manage and store context information; Context Coordinate and Context Association Manager work together to direct the collecting of context information from various sources and the dissemination to clients who issue requests.

The context-aware architecture provides features to preserve personal privacy through interactions between Context Coordinator, Privacy Agent, Privacy Policies Repository, as well as context sources (i.e. human users) and clients (i.e. context-aware applications). The Context Coordinator serves as an interface to context clients, where context information is requested and a basic access control is performed. The basic access control checks if a further fine-grained privacy check by the Privacy Agent in accordance with individuals' privacy preferences is required. Once the privacy check is resolved, an information disclosure agreement between the user and the context client will be stored in the Privacy Policies Repository. Figure 2 illustrates how various components in our architecture work together to preserve privacy when the location-tracking service in our use scenario requests Alice's location information.

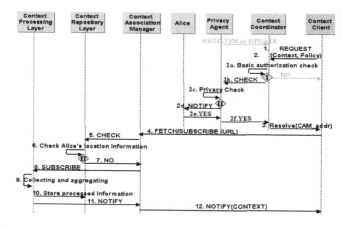

Fig. 2. Sequence of messages that characterizes an authorized context request

4 Privacy Vocabulary and Privacy Rule Ontology

Privacy agent is designed to relieve people from the burden of managing their privacy preferences toward dynamic context-aware environments, in addition to notifying them of relevant information disclosure. It has two major functionalities. On the one hand, it mediates privacy-related interactions between a user and data collectors, including notifying the user of relevant information disclosure and negotiating on behalf of the user with data collectors in accordance with his privacy preferences. On the other hand, the privacy agent serves as a continuously running service that can be contacted and queried by the user anytime, allowing instant access and adjustment to privacy preferences.

To cope with the concern that individuals' privacy preferences might change over time and in response to contexts, some level of automated preference mechanisms (i.e. automatically computing an individual's privacy preferences according to his initial settings) is required, and the privacy agent has an inference engine planted in order to compute disclosure policies in various context. To facilitate the automated processes of the privacy agent, we have been developing a privacy vocabulary to represent privacy data and rules, and using ontological modeling techniques to model the privacy vocabulary. The following subsections introduce the development of the privacy vocabulary and the privacy rule ontology respectively.

4.1 Privacy Vocabulary

Recalling the use scenario of privacy agent, Alice's privacy agent parses and compares the context-aware application's collecting policies against her privacy preferences, negotiates on behalf of Alice with the context-aware application if conflicting interest occurs, and produces a concise report once information disclosure is agreed. The collecting policy of the context-aware application, the privacy preferences set by Alice, and the disclosure agreement are all expressed with a shared set of privacy vocabulary. The privacy vocabulary consists of an unambiguous representation of privacy data, as well as descriptions of disclosure conditions of the privacy data, by which both parties (Alice and the application) and privacy-related functional components involved in our architecture (i.e. Privacy Agent, Privacy Policies Repository, Context Coordinator) could have a common understanding about privacy requirements while interacting with one another.

We have been developing the privacy vocabulary based on the terminology and policies specified in Platform for Privacy Preferences Project (P3P) [1], and adopted P3P policies as a basic data format in privacy data exchanges, with the intention of benefiting from the substantial legal and social expertise that has been put into the development of the P3P standards. However, since the P3P is initially an attempt to

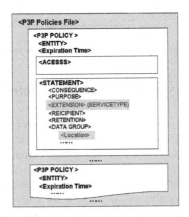

Fig. 3. A high-level skeleton of a P3P policies file (a full explanation about regular P3P policy elements is available in P3P specification [1])

provide privacy mechanisms for Web, it only takes into account a person's identifying information (such as name, birthday, home-address, credit card details, etc.) as private data to be protected. In context-aware environments, staple contextual information (such as a user's location) is also sensitive, but is not covered by the P3P specification. Some extensions are thus necessary to P3P base data schema and regular policy elements before P3P practices could be adopted in context-aware ubiquitous computing environments. In particular, we define a new location data element <Location> to represent a user's current location, and extend P3P's <PURPOSE> element to enable data collectors (i.e. context-aware applications) to explicitly describe their purpose of data collecting practices (in other words, the type of service they offer). Figure 3 below shows a high-level skeleton of the P3P policies file that is used in privacy interactions in our architecture, with two blocks in shadow highlighting the extensions of <Location> and <SERVICETYPE> elements.

4.2 Privacy Rule Ontology

In the field of knowledge management, ontology represents a formal description of concepts in a domain, properties of each concept, and restrictions on those properties, and has inherent strength in capturing relationships between the concepts and properties [10]. This can be used by inference engine planted in privacy agents to reason over ontology descriptions as a means to support privacy check and matching.

We have been experimenting on using ontological modeling techniques to model the privacy vocabulary (including both privacy data elements and disclosure conditions), and attempting to take advantage of existing description logic inference tools to implement ontology-based reasoning. Figure 4 below illustrates a subset of the ontological specification of privacy rules that correspond to P3P specification.

As illustrated in the Figure 4, a Privacy_Rule class is defined to represent privacy preferences set by a person. Every privacy rule is expressed with two elements: Data (Data class) and Conditions (Condition class). The Conditions class contains all conditions under which a person is willing to disclose personal data. According to

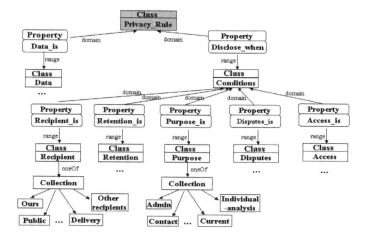

Fig. 4. A subset of the ontological specification of privacy rules

P3P specification [1], the conditions can be classified based on various personal concerns including recipients of data, purposes of data collection, duration that data will be kept by recipients, a user's access privilege to his personal data once stored by recipients, and ways of handling disputes. The Privacy_Rule class has two properties: Data_is and Disclose_when, forming a triple expression that can effectively describe the relationships between privacy rule, data and disclosure conditions.   Both Disclose_when and Data_is are allowed to have multiple values, since a set of data may have same disclosure conditions.

The Data element specified in the Privacy Rule Ontology represents sensitive personal information that asks for privacy protection. The information includes P3P base data scheme and our extensions of location-related contextual information. Since data schema in the P3P specification is structured hierarchically (by using a dotted notation, such as user.home-info.telecom.telephone), it is reasonable to use ontological modeling technologies to capture the multiple-level hierarchy of P3P data scheme. With logic relationships inherent in the ontology-based representation of data scheme, our approach provides some powerful inference capabilities that are not supported by other P3P rule matching languages, such as APPEL [11]. For instance, knowing that a user does not want to reveal her home address and that home telephone number is associated with home address, the privacy agent could reason that it should also keep secret of the user's home phone number.

5   Implementation of Privacy Rule Ontology and Privacy Agent

Privacy Rule Ontology has been developing by using Web Ontology Language (OWL) [12]. HP's Jena platform [13] has been chosen as a programming environment for developing privacy rule inference mechanisms.

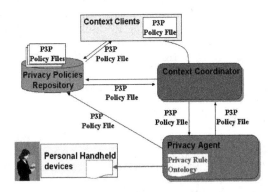

Fig. 5. Privacy Rule Ontology and P3P policy files in privacy-related interactions

As illustrated in Figure 5, the Privacy Rule Ontology (in an OWL file) resides in personal Privacy Agent and is made available to users via their personal handheld devices (such as PDA, Smartphone, etc). The OWL file contains all classes and properties that are required to construct privacy rules, but it does not include rule instances. The rule instances are dynamically generated when users specify manually their privacy preferences. Each time a user edits his privacy preferences, his personal handheld device refers to the Privacy Rule Ontology that is preloaded to the device's memory. Once the new privacy preference is created and sent to the person's Privacy Agent, the Privacy Agent invokes the privacy rule OWL file through Jena API, and creates automatically the instance of Privacy_Rule, Data and Conditions classes, as well as associates the instances with relevant properties.

Privacy agents could be implemented as programming codes and embedded into personal handheld devices or other data management tools, or deployed as a proxy server. In our implementation, we prefer placing privacy agents somewhere on network that is always accessible whenever requested, rather than embedding them into personal handheld devices. The preference is primarily driven by the consideration of saving power and an availability reason. In our proposal, ontology-based reasoning capabilities and a powerful inference engine are required to enable efficient privacy check and rule matching, which probably imposes high requirements on resource-constrained devices like PDA. In addition, in context-aware environments where personal devices might suffer from intermittent connectivity, a remotely located privacy agent could potentially carry out its function independent of the envisioned poor connectivity.

6 Related Work

The development of the privacy agent is an integrated part of our ongoing effort towards developing privacy protection solutions for context-aware systems. During the design of the privacy-respecting context-aware architecture, we had investigated some ubiquitous computing prototypes and systems that were specifically designed with privacy protection in mind, such as Confab [2] by Hong, PawS [7] by Langheinrich, Privacy solutions in AURA project [5] and IETF's Geopriv framework [3]. Our privacy solution has been building upon their experience and attempted to build privacy flavor into the initial architecture design, in order to lessen the risks of providing only shallow and short-lived privacy solution. More importantly, the privacy work proposed in this paper is meant to empower people with appropriate mechanisms to express and manage their privacy preferences with relative ease, which has not been a focus of the privacy work mentioned above.

Applying P3P practices to ubiquitous computing environments has been proposed by [7, 14]. In particular, PawS [7] by Langheinrich presented an informative work that adapted the P3P policies to be applicable in ubiquitous computing environments, which serves as an important supplement and is compatible to our work. However, there is a key difference between our work and other privacy work that has attempted to use P3P. We have been employing P3P terminology and policies, both for data collectors to state collecting policies and for individuals to express privacy preferences. On the contrary, the P3P itself and most of the privacy work built upon the P3P limited the use of P3P policies only as a vehicle for data collectors to state their collecting requirements. They must employ other preference formulation languages, such as APPEL [11], to allow users to express their privacy preferences.

Increasing interest in ontologies in the last few years has led to emerging ontology-based context modeling approaches. Ontology-based context models have been independently developed by several research groups [4,10,15,16]. This trend reflects the potential of ontology-based approaches to address critical issues including formal context representation, knowledge sharing and logic-based reasoning about context. However, unlike context ontologies above (except [16]), which limited the use of ontologies only to represent context information and relationships between context information, we have employed the ontological modeling approach to express privacy vocabulary. By taking advantage of the real power of ontologies as an enabler for logic-based inference, personal privacy agents could have efficient privacy check and matching processes to judge the acceptability of data collectors' collecting policies, therefore taking appropriate actions on behalf of individuals.

7 Conclusions and Future Work

This paper has presented an attempt to develop intelligent agent technologies to enable individuals to manage their privacy requirements toward dynamic context-aware environments with relative ease. The privacy agent approach taken by our work serves as a supplement to privacy protection through conventional access control and security mechanisms.

The development of the privacy vocabulary and ontology presented in this work is among the first step to provision automated preference mechanisms in privacy agents. We are developing a rule-based privacy policy language to be used for expressing and reasoning context-dependent privacy preferences. In addition, we plan to enhance our privacy protection framework by taking into account the deployment of security mechanisms and a trust model in the proposed context-aware architecture.

References

1. Cranor, L., Dobbs, B., Egelman, S., Hogben, G., and Schunter, M.: The Platform for Privacy Preferences 1.1 (P3P1.1), http://www.w3.org/P3P/, July 2005
2. Hong, J.I., and Landay, J.A.: An Architecture for Privacy Sensitive Ubiquitous Computing. In Proceedings of the 2nd international conference on mobile systems, applications, and services (MobiSYS '04), Boston, USA, June 6-9, 2004, ACM Press, (2004) 177–189
3. Cuellar, J., Morris J., Mulligan, D., Peterson, J., and Polk J.: Geopriv Requirement, RFC 3693, IETF, http://www.ietf.org/rfc/rfc3693.txt, February 2004
4. Gandon, F. L., and Sadeh, N. M.: Semantic Web Technologies to Reconcile Privacy and Context Awareness. Web Semantics Journal Vol.1 (3), (2004)
5. Hengartner, U., and Steenkiste, P.: Access Control to Information in Pervasive Computing Environments. In 9th Workshop on Hot Topics in Operating Systems, Hawaii, May 2003
6. Zhang, G. and Parashar, M.: Context-aware Dynamic Access Control for Pervasive Applications. In Proceedings of Communication Network and Distributed Systems Modeling and Simulation Conference, San Diego, USA (2004)
7. Langheinrich, M.: Personal Privacy in Ubiquitous Computing – Tools and System Support. PhD thesis, No. 16100, ETH Zurich, Zurich, Switzerland, May 2005.

8. Hull, R., Kumar, B., Lieuwen D., and Patel-Schneider, P.: Enabling Context-Aware and Privacy-Conscious User Data Sharing, In Proceedings of the IEEE International Conference on Mobile Data Management, Berkeley, USA, January 2004, 187-198.
9. Zhang, N., and Todd, C.: A Generic Context-aware Architecture Prototype, In Proceedings of London Communication Symposium 2005, London, UK, September 2005
10. Wang X. H., Zhang, D. Q., Gu, T., and Pung, H. K.: Ontology Based Context Modeling and Reasoning Using OWL, In Proceedings of the 2nd IEEE Annual Conference on Pervasive Computing and Communications Workshops (PerCom'04), Orlando, USA, March 2004, 18-22
11. Cranor, L., Langheinrich, M., and Marchiori, M.: A P3P Preference Exchange Language 1.0 (APPEL1.0) W3C Working Draft, April 2001, http://www.w3.org/TR/P3P-preferences
12. Smith, M.K., Welty, C., and McGuinness, D. L.: OWL Web Ontology Language Guide, http://www.w3.org/TR/owl-guide/ , February 2004
13. Jena Semantic Web Framework, http://jena.sourceforge.net/
14. Myles, G., Friay, A., and Davies, N.: Preserving Privacy in Environments with Location-based Applications. IEEE Pervasive Computing, Vol.2 (1), (2003) 56-64
15. Henricksen, K., Livingstone, S., and Indulska, J.: Towards a Hybrid Approach to Context Modeling, Reasoning and Interoperation, In Proceedings of the 1st International Workshop on Advanced Context Modeling, Reasoning, and Management in UbiComp'04, Nottingham, England, September 2004
16. Chen, H., Finin, T. and Joshi, A.: Semantic Web in the Context Broker Architecture. In Proceedings of the 2nd IEEE Conference on Pervasive Computing and Communications (PerCom'04), Orlando, Florida, USA, March 2004, 277-286

Ensuring Privacy in Smartcard-Based Payment Systems: A Case Study of Public Metro Transit Systems

Seng-Phil Hong[1] and Sungmin Kang[2]

[1] School of Computer Science & Engineering Sung Shin Women's University, Seoul, Korea
philhong@sungshin.ac.kr
[2] College of Business Administration, Chung-Ang University, Seoul, Korea
smkang@cau.ac.kr

Abstract. The advances in technology have enabled us to share information, process data transactions, and enhance collaborations with relevant entities effectively. Its unparalleled adoption in both the public and private sectors is raising heightened concerns, particularly in the areas of the collection and management of personal information. The use of personal information can provide great benefits, including improved services for customers and increased revenues and decreased costs for businesses. However, it has also raised important issues such as the misuse of their personal information and loss of privacy. In this paper, we propose a framework to preserve privacy in new Public Metro Transit Systems that incorporates smartcard-based payment systems. The proposed framework leverages cryptographic protocols and an innovative privacy model to ensure the protection of privacy information of the cardholders. We also overview our system architecture for the proposed framework including case learned.

1 Introduction

The recent survey indicates that online and offline retailers lost $6.2 billion in sales because of privacy issues [1]. A separate survey found that more than 50% of consumers reported leaving e-commerce sites they have been using because of privacy reasons. These surveys signify that the use of personal information can provide great benefits, including improved services for customers and increased revenues and decreased costs for businesses. However, it has also raised important issues such as the misuse of their personal information and loss of privacy. We have recently witnessed similar issues in new transportation systems in South Korea. One of major cities in South Korea has introduced new metro transportation systems, adopting highly innovative technologies since 2004. They have implemented an overhaul of their public transportation systems, introducing new bus routes, numbers, colors, and adjustments in bus and subway fares. The new transportation systems are designed to achieve both faster and more convenient transit services to the citizens. The new transportation systems have brought many changes and one of important changes is the new type of e-payment smartcard called TP card.

TP card works as an e-cash card, which can be used to pay for transportation fares conveniently, and it can be recharged for the further usage. Since TP card has an

H. Leitold and E. Markatos (Eds.): CMS 2006, LNCS 4237, pp. 206–215, 2006.
© IFIP International Federation for Information Processing 2006

integrated circuit chip, it can not only be used as an ID card and pay for transit fares, but also be used to make small purchases at the convenient stores, restaurants, and movie theaters. More specifically, it is a smartcard with the improved security technologies based on domestic security technology standards and international security algorithms. Application areas of TP can be extended to e-payment for other transportation fares and e-ticketing for product and service purchases in the near future.

With 12 months of its usage in effect, there are advantages in newly adopted Public Metro Transit System, called e-PTS, and some early problems are encountered with challenges of requirements to resolve the inconvenience of new e-payment systems and technical threats related to security and privacy issues. In this paper, we identify the problems that exist with the advent of e-PTS and suggest countermeasures to resolve such technical problems. A recent case of criminal mischief led us to consider the above-mentioned issues. The law enforcement agency caught the criminals by tracing TP card usage information in e-PTS. However, this is an important wake-up call to other users because it can be a critical problem to the protection of their privacy. Therefore, no trace at all or only limited trace of personal information should be allowed on payment methods, providing assurance level of anonymity. This prompts the need for more secure method of transportation fare payment. Limited tracing of personal information needs to be implemented with the cooperation of few designated agencies. The goal of this work is to suggest and analyze the practical privacy model to the existing e-PTS.

The rest of this paper is organized as follows. Section 2 discusses background technologies followed by the overview of e-PTS in Section 3. In Section 4, we propose a privacy model for e-PTS including system architecture. Section 5 describes features of the proposed model. Section 6 concludes this paper.

2 Background Technologies

2.1 Public Key Infrastructure (PKI) and Digital Signature

PKI is an infrastructure for disseminating the public key in a secure and reliable channel. One of important components of PKI is a set of certificate authorities (CAs) that archives public keys of certified users or entities. The user or entity that wishes to participate in this infrastructure must successfully prove their identity to the CA [2, 3, 4]. Even though some argued the risks on security services of PKI [5], PKI has been considered as a viable solution for security and privacy services by healthcare industries. Hence, our work utilizes PKI to develop a scalable privacy model for e-payment system.

2.2 Smart Token Technologies

Smart tokens are devices with a memory and a processor which can generate and store keys. It also supports cryptographic functions such as encryption, digital signature, or key agreement. Some noticeable characteristics of smart tokens are portability, tamper-resistant storage, and isolation of computational activities (i.e.

leveraging the features of cryptographic functions without revealing private keys to other system components) [6].

Smartcard. The smartcard, an intelligent token, is a plastic card embedded with an integrated circuit chip. It provides not only memory capacity, but computational capability as well. The security features of smartcard make it resistant to security threats. A smartcard is a card that is embedded with either a microprocessor and a memory chip or only a memory chip with non-programmable logic. The microprocessor card can add, delete, and manipulate information on the card, while a memory-chip card can only undertake a pre-defined operation. [7, 8].

2.3 Threshold Cryptography

The idea of threshold cryptography is to protect information (or computation) by fault-tolerantly distributing it among a cluster of cooperating computers [9, 10, 11]. First consider the fundamental problem of threshold cryptography, a problem of secure sharing of a secret. A secret sharing scheme allows one to distribute a piece of secret information among several servers in a way that meets the following requirements: (1) no group of corrupt servers (smaller than a given threshold) can figure out what the secret is, even if they cooperate; (2) when it becomes necessary that the secret information be reconstructed, a large enough number of servers (a number larger than the above threshold) can always do it.

3 e-PTS: Architecture and Privacy Issues

The e-PTS consists of three major components: Sub, Main, and Linkage systems. Sub, Main, and Linkage systems are interconnected through enterprise application integration (EAI) interface. Sub system components perform activities related to

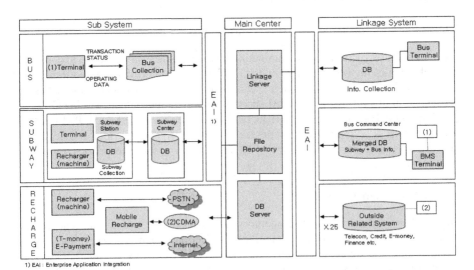

Fig. 1. e-PTS Architecture

re/charging the transportation fare for bus and subway systems. Main center components store information gathered from sub system components and manage and administer the collected information efficiently. Further, Linkage system relays the processed information to the relevant sites, which require metro and e-payment information gathering. Figure 1 illustrates the architecture of e-PTS.

With the introduction of new Public Metro Transit Systems, we have witnessed critical privacy concerns that need to be studied and analyzed to investigate relevant countermeasures. Our study indicated that some of privacy issues are still raising inevitable business problems as follows:

Anonymity: It enables users to make use of e-PTS without being tracked and keep the state of being anonymous or virtually invisible. A user could spend all day using e-PTS but the sites or location information the user visits should be protected.

Pseudonymity: Like characteristics of anonymity, a user cannot be identifiable but the user can be tracked through an alias or persona that the user has adopted.

Unlinkability: It refers to the inability to link pieces of related information. This could mean isolating multiple transactions made using the same TP card. The ability to link transactions could reveal an idea of daily routes or how much expenses have been consumed over a month.

Unobservability: This refers to the inability to observe (or track) while a user is accessing a service. The multi-purpose TP card can be used to abuse payment information without cardholder's permission or legitimate access when the TP card is used for other services. However, it may be useful to observe activities of the user under certain circumstances such as disaster or medical emergency.

Authorization privacy: To recharge a TP card, users are often required to present their identifications. This can be used to track how often a user recharges the TP card, even though it is important only to know that the bearer deserves access to the facility.

Data management: Collected and managed data information can be often misused with malicious intent or by mistakes.

In the subsequent sections, we attempt to articulate possible solutions for key issues involved with the above privacy concerns.

4   Privacy Model for e-PTS

In this section, we propose a policy model for e-PTS, called Privacy Model for Public Metro Transit System (PMPTM). In our model, we focus on the following privacy issues that need to be solved in large-scale distributed environment: when a cardholder conducts multiple smartcard transactions at different places, the cardholder's personal information, transaction data, and critical payment information could be revealed without cardholder's recognition at each location of card transaction.

In our model, we seek mechanisms to help reveal personal information only to the authorized users/entities and enforce privacy policies which are specified by and assigned to the cardholders.

PMPTM architecture is illustrated in Figure 2. It consists of two important components: Privacy Check Box and Policy Bank. The functionalities of each component are as follows:

Privacy Check Box (PCB). PCB is a module for maintaining the confidentiality of a cardholder's information and checking his/her privacy conditions concerned with PMPTM. After a session is established through PMPTM, authentication unit not only checks the cardholder's identification but also determines the cardholder's critical information that is not available for anyone by using a threshold cryptographic protocol. PCB then verifies and validates the cardholder's privacy condition within privacy condition box.

Policy Bank (PB). PB is a set of access control policies for passing a system's request to another system between cardholders and card managers. The PMPTM defines the policy framework consisting of policy gathering component, policy repositories for the assigned policy, and policy enforcement for handling policy decisions. Policies are rules governing the choices in behavior of a system.

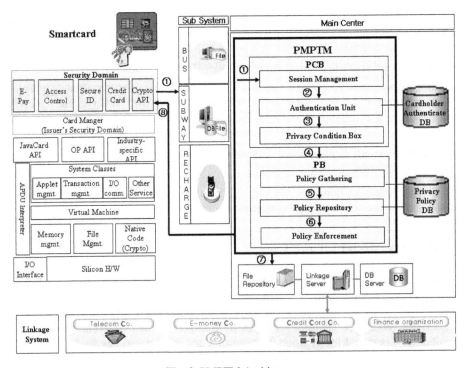

Fig. 2. PMPTM Architecture

Privacy Model for Public Metro Transit System (PMPTM) works as shown in the following process. Once a smartcard holder attempts to access PMPTM mechanism (1), SSL connection is established and Session Management provided to the cardholder's credentials are passed to Authentication Unit (2). Authentication Unit accesses PCB to authenticate the cardholder and PCB (Privacy Condition Box) verifies the cardholder's privacy condition related to privacy information which maintains usernames, privacy policy type, level and principle of privacy policy actions (3). PR (Privacy Repository) gathers privacy information if the cardholder's privacy condition checking is successfully completed. PR retrieves or gathers privacy policy information from private policy database and generates privacy information if it needs to update (4, 5). After PR stores the information and PE (Policy Enforcement) takes an action that provide the well-define access control according to privacy policy (6), PMPTM then triggers the requested service and sends the signed crypto API to the cardholder's smartcard for further transactions and file repository (7, 8). Finally the requested service is provided to the user from PMPTM Service.

In our model, policy types are categorized as one of the following access control policies, which depend on what activities a subject can perform on a set of target objects:

- Policy_Type_A/A′ : A set of subjects must do / not perform a set of target objects.
- Policy_Type_B   : A set of subjects must validate the conflict of access control policies to a set of target objects.
- Policy_Type_C/C′ : Actions are permitted to / prohibited from a set of subjects or a set of target objects
- Policy_Type_D   : Actions are delegated to a set of subjects or a set of target objects.

Table 1. Notation in PMPTM

A (subject)	A principal or system that requests an action
B (object)	A principal or system that can perform an action requested by A
Action	A list of operations: request/response/check/order/cancel/pay/debt
t_1	Time that an input call starts
t_2	Time that an output call starts
r_1, r_2	Random numbers
$cert_A$	A's public key certificate
$cert_{PM}$	PMPTM's public key certificate
pc_A	A's privacy condition attributes. It would be representing to general information and other information such as payment information and payment profiles.
P	Priority information. It has three attributes: low/medium/high
σ_A	Signature of Cardholder A
σ_{PM}	Signature of PMPTM

Policy gathering and policy repository components should guarantee to determine correct access constraints and appropriate policy management including modification and revocation of policies. The policy enforcement should trigger an action to retrieve policies from policy repository and evaluate the policy condition when policy is executed from policy repository.

Next, we describe how each component in PMPTM works to support the proposed architecture. We use notations in Table 1 to explain the details. The PMPTM architecture works as follows:

1. Input Call
   - A cardholder "A" sends a message to the PMPTM making an input call using contactless smartcard based on ISO14443 Type A/B format. The cardholder requests the privacy information to all registered manufacturers in PMPTM.
     - $Input\_call =< A, B, t_1, r_1, request, \sigma_A, cert_A, pc_A >$

2. Policy Check Box
   - Session Management: A user inputs information for establishing a session in PMPTM.
   - Authentication Unit: User Interface sends a session establishment input to PCB. Secure connection (such as X.25) is established and the cardholder's credentials (such as X.509 Certificates) are passed to PCB. And it forwards the cardholder's information to authentication unit then verifies the cardholder's signature using the associated public key.
   - Privacy Condition Box: Authentication unit transmits the cardholder's authentication information to PCB if the authentication check is successfully completed. Then it checks the validation of the cardholder's privacy condition within PCB. Reference monitor transmits the cardholder's information to the PB if the validation check is successfully completed.
   - After all, PMPTM checks the information which request from the cardholder A
     - Check the validation of $cert_A$ using PKI.
     - Verify the signature σ_A in an input call using $cert_A$.
     - Check pc_A.

3. Policy Bank
   - Policy Gathering: It collects the cardholder's information from PCB, then retrieves privileges granted to the corresponding user's action
   - Policy Repository: It assigns reasonable policies that provide a set of cardholder's information from policy repository. It includes action, priority, and policy.
   - Policy Enforcement: It validates and enforces a selected policy.
     - If pc_A is valid, retrieve a privacy information from pc_A.
     - PB executes the policy gathering from the policy repository (as shown in Table 2) and the policy enforcement procedures.

Table 2. Cardholder Domain Table in Policy Repository

SUBJECT	ACTION	Policy TYPE	PRIORITY	OBJECT
Cardholder A	REQUEST	Policy_Type_B	Middle	Card Manager B (PMPTM)
Cardholder A	PAY	Policy_Type_C	High	Card Manager B (PMPTM)

4. Output Call
- Finally, a reformatted message is sent to a set of target objects or returns to cardholders. PMPTM generates output call for each registered card manager B in PMPTM

  ○ $Output\_call = <A, B, t_2, r_2, pc_A, P, \sigma_{PM}, cert_{PM}, >$

  Where pc_A is general information and P is middle.
- PMPTM resends output call to cardholder A. The procedure of response is the same as the request procedure. The cardholder A can gather the authentication information, privacy condition and price from each response. The requests such as "order" and "pay" can be executed through PMPTM mechanism.

5 Features of PMPTM

This section describes the strength of PMPTM in comparison with previously discussed problem statement in section 3.

Strong Privacy Control: Among the various functions of smartcard service, information leakage should be prevented, as a cardholder needs to protect his or her privacy. Using the threshold cryptography protocol, the card holder's personal and transaction information, which can be exposed to others by service provider without cardholder's consent, should be informed to the cardholder first before sharing the information. Moreover, we tried to resolve the "privacy issues" in the large-scale distributed environment by suggesting the efficient access control measures in the PMPTM architecture. Through these access control measures, we emphasize the importance of data confidentiality and data integrity that have been overlooked previously due to data traceability and data availability. We believe data confidentiality and data integrity should be considered as more important security requirements from the privacy perspective.

Well-defined Privacy Policy: A cardholder wishing to access community resources contacts the PMPTM server. The card manager can easily keep track of cardholder's privacy condition and select fine-grained access control policies, because PCB mechanism can verify and validate cardholder's personal information. In PB, it also provides centralized monitoring condition, which handles how well card holders can present themselves to gain access on behalf of the business community condition. It could improve ease of use and accuracy of the administration process even if access control is implemented in a variety of heterogeneous components, and the cardholder or card manager needs to concentrate only on this very unit in which all security related configurations are maintained.

Trust Policy Management: The PMPTM mechanism should rely on a simple mode transaction that covers the different range of large-scale multi-organizational systems. Because PMPTM mechanism strictly enforces stored access policies that have been already defined in PB repository of PMPTM. This means that it is very hard to modify/delete our reserved rules or policies by unauthorized people. Additionally, our policies are supposed to reserve a couple of business action (Request/Response/Check/Verify/Pay/Debt) cases. There is no change in the policy type without permission or privilege.

Scalability and Performance: When cardholder sends the message to PMPTM, PMPTM checks user's status (subject, object, privacy condition, action, and security level). The priority information of message represents how important the business condition is. Practically, it is very helpful to control many users who have different security conditions simultaneously. If card manager finds the priority message (Security level-H: High), then they should more carefully execute the message such as dual verification and end-to-end monitoring, or the high level security message should be sent first if other messages/queues are waiting for next processing. This could enable congestion control. The cost of adding or removing participants, which corresponds to changing policy types, should not increase the number of resource providers participating in PMPTM. Therefore, resource administration overheads should also be controlled and minimized.

6 Conclusion and Future Work

The new transportation card system based on smartcard technology shifted a paradigm of public service from offline environment to online environment. We attempted to implement new IT infrastructure for automatic fare collection mechanisms and efficient management systems based on the use of smartcard in the distributed computing environment. In this paper, we proposed a new privacy model and architecture, which can be easily implemented in order to resolve the security and privacy issues that exist with respect to the protection of personal information and privacy. We have also described lessons learned through major features in e-PTS so that system engineers and software developers can adopt our approach to implement the relevant system. We believe our work will help facilitate the growth of e-payment service based on TP in e-PTS. As a future research direction, we plan to conduct a research by considering the extension of the suggested model and smartcard management using the notion of role-based access control effectively.

References

1. A list of privacy surveys, Available at http://www.privacyexchange.org/iss/surveys/surveys.html
2. Min Liu, Shudong Sun, and Miaotiao Xing, "Study on security based on PKI for e-commerce of statistics information system", ACM International Conference Proceeding Series; Vol. 113, ACM Press, Xi'an, China, Aug. 2005, pp. 729 – 732.

3. L. Pearlman, V. Welch, I. Foster, C. Kesselman, S. Tuecke, "A Community Authorization Service for Group Collaboration Policies for Distributed Systems and Networks," Proceedings of the Third International Workshop in 2002, pp. 50–59.

4. Pierangela Samarati, Michael K. Reiter, and Sushil Jajodia, "An Authorization Model for a Public Key Management Service," ACM Transactions on Information and System Security, 4(4), Nov. 2001, pp. 453-482.

5. C. Ellison and B. Schneier. Ten Risks of PKI: What you are not being told about Public Key Infrastructure. *Computer Security Journal*, 16(1): 1-7, 2000.

6. Xinhua Zhang, Christoph Meinel, and Alexandre Dulaunoy, "A Security Improved OpenSST Prototype Combining with SmartCard," Proceeding of the International Conference on Computer Networks and Mobile Computing, IEEE, 2003.

7. Blerim Rexha, "Increasing User Privacy in Online Transactions with X.509 v3 Certificate Private Extensions and Smartcards", Proceedings of the IEEE International Conference on E-Commerce Technology, Washington, USA, July, 2005, pp. 293–300.

8. Yankiang Yang, Xiaoxi Han, Feng Bao, Deng R.H., "A Smart-card Enabled Privacy Preserving E-prescription System," IEEE Transaction on Information Technology in Biomedicine, Vol. 8, No. 1, pp. 47-58.

9. A. Shamir, "How to share a secret," Communication of the ACM, Vol. 22, No.11, pp. 612-613.

10. Anna Lysyanskaya, Chris Peikert, "Adaptive Security in the Threshold Setting: From Cryptosystems to Signature Schemes," ASIACRYPT, 2001.

11. Ran Canetti, Shafi Goldwasser, "An Efficient Threshold Public-key Cryptosystem Secure Against Adaptive Chosen Ciphertext Attack," EUROCRYPT, 1999, pp. 90-106.

Attack Graph Based Evaluation of Network Security

Igor Kotenko and Mikhail Stepashkin

SPIIRAS, 39, 14 Liniya, St.-Petersburg, 199178, Russia
{ivkote, stepashkin}@comsec.spb.ru

Abstract. The perspective directions in evaluating network security are simulating possible malefactor's actions, building the representation of these actions as attack graphs (trees, nets), the subsequent checking of various properties of these graphs, and determining security metrics which can explain possible ways to increase security level. The paper suggests a new approach to security evaluation based on comprehensive simulation of malefactor's actions, construction of attack graphs and computation of different security metrics. The approach is intended for using both at design and exploitation stages of computer networks. The implemented software system is described, and the examples of experiments for analysis of network security level are considered.

Keywords: Network security, Vulnerability Assessment, Risk Assessment, Security Metrics, Network attacks

1 Introduction

The increase of networks and security mechanisms complexity, vulnerabilities and potential operation errors as well as malefactors' possibilities causes the necessity to develop and use powerful *automated security analysis techniques*. These techniques should allow revealing possible assault actions, determining vulnerabilities, critical network resources and security bottlenecks, and finding out and correcting errors in network configurations and security policies.

At *design stages*, the different approaches to security analysis can be used, for example, based on qualitative and quantitative risk analysis. Approaches based on building the representation of malefactor's actions in the form of attack trees or attack graphs, the subsequent checking of various properties of these trees or graphs on the basis of usage of different methods (for example, model checking), and determining various security metrics are the perspective directions in evaluating security level of large-scaled networks. At *exploitation stages*, passive and active methods of vulnerability assessment are used. The passive methods do not allow estimating the possible routes of malefactor's penetration. The active methods can not be applied in all situations, as lead to operability violation of network services or the system as a whole. The combination of passive methods (for obtaining appropriate data about network configuration and security policy), procedures of attack graph construction, and automatic reasoning allows solving partially these two problems.

The paper is devoted to creating the architecture, models and prototypes of security analysis system (SAS) based on construction of attack graphs and computation of different security metrics on the basis of combination of qualitative risk analysis

H. Leitold and E. Markatos (Eds.): CMS 2006, LNCS 4237, pp. 216–227, 2006.

techniques. SAS is based on the following functions: simulating malefactor's activity; building possible assault actions graph; analyzing malefactors' actions from different network points and directed to implementing various security threats; revealing vulnerabilities and security "weak places" (the most critical computer network components); calculating different security metrics and evaluating general security level; comparison of retrieved metrics and user requirements and elaboration of recommendations on security increase. The work is organized in the following way. *Section 2* is an overview of relevant works and the suggested approach peculiarities. *Section 3* represents the model of attack scenarios and the common attack graph generated. *Section 4* specifies security metrics and main phases of evaluating a general security level. *Section 5* emphasizes the approach complexity problems and solutions. *Section 6* considers the generalized architecture of SAS. *Section 7* describes the examples of experiments fulfilled with SAS. *Conclusion* surveys the results and further research.

2   Related Work and the Approach Peculiarities

There are a lot of works which consider *various approaches to security analysis.*

Alberts and Dorofee [1] as well as Chapman and Ward [2] described different risk analysis techniques for estimating security level. Ritchey and Ammann [14] proposed model checking technique for network vulnerability analysis. Jha et al. [6] suggested the technique of attack graph evaluation based on model checking and probabilistic analysis. Sheyner et al. [16] presented algorithms for generating scenario graphs based on symbolic and explicit-state model checking. These algorithms ensure producing counterexamples for determining safety and liveness properties. Rothmaier and Krumm [15] suggested an approach for analyzing different attack scenarios based on a high-level specification language, a translation from this language to constructs of model checker, applying optimization techniques and model checking for automated attack scenario analysis.

Lye and Wing [7] suggested the security evaluation method based on game theory. The authors view the interactions between an attacker and the administrator as a two-player stochastic game and construct the game model. The approach offered by Singh et al. in [17] is intended for performing penetration testing of formal models of networked systems for estimating security metrics. Swiler et al. [18] proposed an approach for construction of attack graph which can identify the attack paths with the highest probability of success. Hariri [5] described global metrics which can be used to analyze and proactively manage the effects of complex network faults and attacks, and recover accordingly. Rieke [13] offered a methodology and a tool for vulnerability analysis which can automatically compute attack paths and verify some security properties. Dantu et al. [4] proposed an approach to estimate the risk level of critical network resources using behavior based attack graphs and Bayesian technique. Ou et al. [12] suggested a logic programming approach to automatically fulfill network vulnerability analysis. Noel and Jajodia [8] considered the common approach, attack graph visualization techniques and the tool for topological vulnerability analysis. Ning et al. [10] suggested different techniques to construct high-level attack scenarios.

The paper suggests *a new approach to security evaluation based on comprehensive simulation of malefactor's actions, construction of attack graphs and computation of different security metrics.* The main differences of offered approach from examined ones consist in *the way of modeling assault actions* (we use a multi-level model of attack scenarios) and applying constructed attack graphs (for different locations of malefactors) *to determine a family of security metrics and to evaluate different security properties.* While the first feature has been taken into account partly in previous works, the second one mainly has not been considered. The third peculiarity of offered approach is that it can be used at different stages of computer network life cycle, including design and exploitation stages. At design stage, the Security Analysis System (SAS) founded on this approach should use the given specifications of computer network and security policy. At exploitation stage, it interacts with a real computer network getting the necessary configuration and policy data in passive mode. The results of security analysis are vulnerabilities detected, attacks routes (graphs), network bottlenecks, security metrics, which can be used for general security level evaluation of network and its components. Obtained results allow producing the valid recommendations for eliminating detected vulnerabilities and bottlenecks, as well as strengthening the network security level.

3   Attack Scenarios and Generalized Attack Graph

Generalized attack scenario model is hierarchical and contains three levels: integrated level, script level and action level. The *integrated level* determines high-level purposes of the security analysis directed to main security threats realization and analyzed objects (particular hosts, network fragments or the whole network). Integrated level allows coordinating of several scenarios. These scenarios may be performed by both one malefactor and malefactors group. The *script level* takes into account malefactor's skill and initial knowledge about network, defines attack object and purpose (for example, "host OS determining", "denial of service", etc.). Script level contains a set of script stages and substages. The main stages are reconnaissance, penetration (initial access to the host), privileges escalation, threat realization, traces hiding, backdoors creation. The *action level* describes low-level malefactor's actions and exploits.

The algorithm of generating the common attack graph is intended for building the attack graph which describes all possible routes of attack actions in view of malefactor's initial position, skill level, network configuration and used security policy. The algorithm is based on the action sequence set in the attack scenarios model: actions which are intended for malefactor's movement from one host onto another; reconnaissance actions for detection of "live" hosts; reconnaissance actions for detected hosts; attack actions based on vulnerabilities and actions of ordinary users.

All *objects of general attack graph* are divided into two groups: base (elementary) objects and combined objects. *Base objects* define the graph vertexes. They are linked by edges for forming the different sequences of malefactor's actions. *Combined objects* are built on the basis of linking the elementary objects by arcs. Objects of types "host" and "attack action" are base (elementary) objects. Objects of the types "route", "threat" and "graph" are *combined objects*. *Route* of attack is a collection of linked vertexes of general attack graph (hosts and attack actions), first of which represents a host (initial malefactor's position) and last has no outgoing arcs. *Threat* is a set of various attack routes having identical initial and final vertexes. Classification of attack actions allows differentiating threats as *primary threats* (confidentiality, integrity and availability violation) and *additional threats* (gaining information about host or network, gaining privileges of local user or administrator).

4 Security Level Evaluation

Determining each security metric and the general security level of analyzed network can be realized in different ways. We use two approaches for security level evaluation: Qualitative express assessment on basis of qualitative methodologies of risk analysis; Quantitative computation of network security level (on basis of Bayesian networks, possibility theory and fuzzy sets). This paper presents the first approach.

The set of security metrics was constructed on basis of general attack graph. Security metrics describe security of both base objects and complex objects of general attack graph. *Examples of security metrics* are as follows: (1) *Metrics based on network configuration* (Quantity of hosts, firewalls, Linux hosts, Microsoft Windows hosts, hosts with antivirus software installed, hosts with personal firewalls, hosts with host-based intrusion detection systems, etc.); (2) *Metrics of hosts* (Criticality level, etc.); (3) *Metrics of attack actions* (Criticality level; Damage level; Access complexity; Base Score; Confidentiality Impact; Availability Impact; Access Complexity, etc.); (4) *Metrics of attack routes* (Route length expressed in vulnerable hosts; Route average Base Score; Maximum Access Complexity; Damage level of route; Maximum damage level of route, etc.); (5) *Metrics of threats* (Minimum and maximum quantity of different vulnerable hosts used for threat realization; Quantity of different routes which lead to threat realization; Damage level of threat; Maximum damage level of threat; Access Complexity of threat; Admissibility of threat realization; Risk level of threat, etc.); (6) *Metrics of common attack graph* (Quantity of different vulnerable hosts of graph; Quantity and set of different attack actions, Average Base Score of all different attack actions, Quantity of routes leading to confidentiality, integrity, availability violations, Quantity of treats leading to confidentiality, integrity, availability violations, Integral security metric "Security level", etc.).

Some security metrics are calculated on basis of standard *Common Vulnerability Scoring System* [3]. CVSS metrics are divided into three main groups: *Base indexes* define *criticality* of vulnerability (attack action realizing given vulnerability); *Temporal indexes – urgency* of the vulnerability at the given point of time; *Environmental indexes* should be used by organizations for priorities arrangement at time of generating plans of vulnerabilities elimination.

The offered approach of qualitative express assessment of network security level contains the following stages: (1) Calculating the criticality level of hosts (*Criticality(h)*, $\forall h \in [1, N_H]$, N_H – hosts amount) using three-level scale (*High, Medium, Low*); (2) Estimating the criticality level of attack actions (*Severity(a)*, $\forall a \in [1, N_A]$, N_A – actions amount) using the CVSS algorithm of action criticality assessment; (3) Calculating the damage level of attack actions (*Mortality(a,h)*, $\forall h \in [1, N_H]$, $\forall a \in [1, N_A]$) taking into account criticality levels of actions and hosts; (4) Determining the damage level of all threats $(Mortality(T)=Mortality(a_T,h_T)$, $\forall T \in [1,N_T]$, where N_T − threats amount, a_T − latest attack action directed on the host h_T for threat T); (5) Calculating the metrics of "Access complexity" for all attack actions $(AccessComplexity(a)$, $\forall a \in [1,N_A])$, all routes $(AccessComplexity(S)$, $\forall S \in [1,N_S]$, N_S − routes amount), and all threats $(AccessComplexity(T)$, $\forall T \in [1,N_T])$; (6) Estimating the admissibility of threats realization $(Realization(T)$, $\forall T \in [1,N_T])$ using the metrics of "Access complexity"; (7) Network security level $(SecurityLevel)$ evaluation using the estimations of threats admissibility and damage level caused by threats realization. Four security levels are used: Green, Yellow, Orange and Red.

5   Complexity Problems and Solutions

The complexity of generating the attack graph is determined by the quantity of malefactor's actions. The given quantity depends mainly on the quantity of hosts in analyzed network (N_H) and the quantity of used vulnerabilities (exploits) from the internal database of vulnerabilities (N_V).

Let us consider the test network which includes n hosts. Each of these hosts has vulnerabilities allowing malefactor to gain a root privileges on the host and to move to the compromised host to attack others. During scanning process the malefactor can reveal all n hosts and realize all attack actions and movements to the captured hosts. Therefore the following formula can be used to approximately compute the complexity of generating the attack graph:

$$F(N_H,N_V) = N_H N_V F(N_H-1,N_V) = N_H N_V (N_H-1) N_V F(N_H-2,N_V) = N_V^{N_H} N_H!$$

The complexity of attack graph analysis is determined by the complexity of attack graph depth-first traversal and is equal to $O(V + E)$, where V − graph vertexes; E − graph edges.

This discussion shows that the given approach faces a combinatorial explosion in complexity. So, it can be applied with success to small networks, but cannot be used without corresponding modification for large scaled networks.

The following *approaches for reducing the complexity* of generating the attack graph are suggested:

1. Splitting the network into fragments, parallel computing for each fragment with subsequent combination of results.

2. Aggregation and abstraction of representations of attack actions and (or) network objects:

- *Using attack actions.* The type of aggregation or abstraction is selected according to offered generalized attack scenario model. The examples of attack actions aggregation objects are as follows: main stages (reconnaissance, penetration, privileges escalation, etc.), attack class (buffer overflow, DoS, etc.), attack subclass (for example, specific type of buffer overflow). Thus, at attack graph construction it is possible to merge a set of vertexes representing actions of one type in one vertex (for example, actions "ping", "get banners", "get services" can be merged into a set of actions named "reconnaissance").

- *Using network objects.* The combination of several hosts of network segment, network segment, the combination of network segments can be aggregated network objects. Thus, the analyzed network can be splitted into aggregated objects which are represented as one object. Such object can be characterized by a set of parameters inherent to all its components (hosts, routers, network switches). The list of operating systems types, the list of network services, etc. can be used as elements of such set of parameters. The reduction of attack graph complexity in such case is achieved because instead of a big set of vertexes (where the hosts are targets of attack actions), the significantly smaller set of vertexes (where the aggregated objects are targets of attack actions) are displayed on the graph.

- Combining various approaches to aggregation and abstraction.
3. Combining parallel computing, aggregation and abstracting.

In further work the development of various algorithms for generating the attack graph is supposed. These algorithms will differ by accuracy and complexity.

6 Security Analysis System Architecture

The architecture of Security Analysis System (SAS) is depicted in fig.1.

User interface provides the user with ability to control all components of SAS, set the input data, inspect reports, etc. *Generator of network and security policy internal model* converts the information about network configuration and security policy into internal representation. The input information is received from Information collector (at exploitation stage) or from specifications expressed in System Description Language (SDL) and Security Policy Language (SPL) (at design stage). These specifications should describe network components and security with the necessary degree of detail – the used software (in the form of names and versions) should be set. *Data controller* is used for detection of incorrect or undefined data.

The network configuration and security policy database contains information on network configuration and security policy rules (this part is used for generating attack action results) as well as malefactor's view of network and security policy (it is generated as the results of attack actions). It is possible to plan the sequence of malefactor's actions on basis of this database (for example, if malefactor has user privileges and needs to read a file F, and according to security policy only local administrators can read this file, then malefactor must do actions to gain the administrator privileges).

Actions database includes the rules of "IF-THEN" type determining different malefactor's operations. IF-part of each rule contains action goal and (or) conditions. The condition is compared with the data from network configuration and security policy database. THEN-part contains the name of action which can be applied and (or) the link on exploit and post-condition which determines a change of network state (impact on an attacked object). *Actions which use vulnerabilities* (unlike other bases of the given group) are constructed automatically on basis of external vulnerabilities database OSVDB [11]. *Common actions* contain actions which are executed according to user's privileges (for example, "file read", "file copy", "file delete", etc.). Databases of reconnaissance and common actions are created by experts.

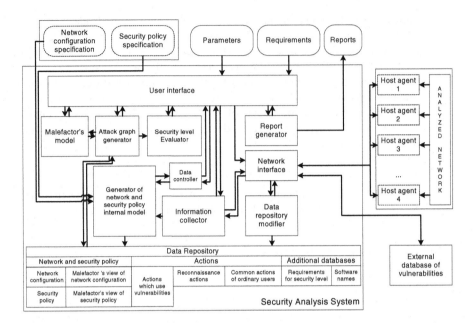

Fig. 1. Architecture of security analysis system

DB of requirements contains the predefined sets of security metrics values (set by experts). Each set corresponds to the certain security class regulated by international standards or other normative documents. The *database of software names* is used by *Data controller* for detection of errors in the specifications of computer network (e.g. when user writes "Orakle" instead of "Oracle") and for generating recommendations on using software tools. In case of detecting discrepancy the conflict is resolved by choosing the correct software name from the list suggested.

Data repository modifier downloads the open vulnerability databases (we use OSVDB [11]) and translates them into actions database. *Attack graph generator* builds attack graph by modeling malefactor's actions using information about network configuration, security policy and available actions from data repository. This module sets up security metrics of elementary objects. On basis of these metrics *Security level evaluator* calculates the metrics of combined objects, evaluates security level, compares obtained results with requirements, finds bottlenecks, and generates recommendations on strengthening security level. *Malefactor's model* determines malefactor's skill level, his initial position and knowledge on network. Malefactor's skill level determines the attack strategy and the set of actions used by malefactor.

Hosts agents serve for passive data gathering. On the basis of these data the network and security policy internal model is formed at exploitation stage. For example, the agents can make the analysis of configuration files of operating system and other software components. *Network interface* provides interaction with external environment (sending requests to external vulnerabilities databases for updates and communicating with agents). *Information collector* interacts with *host agents* and receives from them information about network devices and settings of software components.

7 Experiments

Fig. 2 shows the structure of test computer network used in experiments.

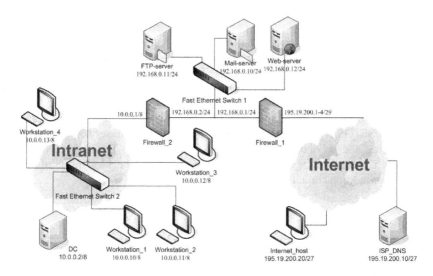

Fig. 2. Structure of test network

When user is working with SAS, he needs to perform the following operations: (1) Loading network configuration and security policy; (2) Setting security requirements; (3) Choosing the high-level purpose of security analysis process; (4) Setting the values of parameters of security analysis; (5) Security analysis process; (6) Modification of network configuration and security policy (if needed).

Network specification in specialized language (SDL) allows defining network topology, information about operating systems of the network hosts, TCP/IP protocol stack settings, services, etc. Security policy description in the specialized language (PDL) allows specifying the network traffic filtration rules for boundary hosts, confidence relations, authentication and authorization rules, etc. Network traffic filtration rules are specified as table collection [9]. Port forwarding rules are specified by tables PREROUTING and FORWARD (all incoming connections not described in the table are forbidden). Table 1 contains notation for the main elements of attack graph.

Table 1. Attack graph elements notation

Malefactor action	Malefactor action (severity and access complexity of action are in brackets)	Malefactor location PRIVILEGES	Malefactor's location and privileges
Final malefactor action	Final malefactor's action	Attacked host (criticality)	Attacked host and its criticality level (in brackets)

Let us consider how SAS works at design stage. Let input data for security analysis are as follows: (1) port forwarding rules for host Firewall_1 are in Table 2; (3) Firewall_1 and Firewall_2 trust to all DMZ hosts;   (4) malefactor is located at external network at host Malefactor, and has administrator privileges; (5) security analysis purpose is to analyze all kinds of threats (integrity, availability, confidentiality violation); (6) security analysis task is to analyze all DMZ and LAN hosts; (7) requirements to analyzed network: the given network should have security level better than Orange. Fig. 3 shows general attack graph for example 1.

Let us consider shortly the process of building the attack graph. At first malefactor is located at the "Malefactor" host and performs 'Ping Hosts" attack. The attack allows him to determine live hosts. Malefactor receives data about four hosts (FTP_server, Web_server, Mail_server and Firewall_1) with IP 195.19.200.1-4 (actually this is only Firewall_1, but malefactor does not know it). Then malefactor analyzes every host separately. Let us consider analysis of the host with IP 195.19.200.2. Four reconnaissance scripts are generated: (1) "Nmap serv" (open port scanning); (2) "Nmap OS" (OS type and version determining); (3) "Nmap serv"+"Banner" (open ports scanning and services identifying); (4) "Nmap serv" +"Banner" +"Nmap OS". After every reconnaissance script realization, malefactor checks if host information satisfies the conditions of actions that use vulnerabilities.

Table 2. Port forwarding rules for host Firewall_1

Comment	Destination		Forward to...	
	IP	Port	IP	Port
Web_server	195.19.200.3	80	192.168.0.12	80
FTP_server	195.19.200.2	21	192.168.0.11	21
MAIL_server POP3	195.19.200.4	110	192.168.0.10	110
MAIL_server SMTP	195.19.200.4	25	192.168.0.10	25
MAIL_server RDC	195.19.200.4	3389	192.168.0.10	3389

The result of the "Nmap serv" action for host with IP 195.19.200.2 is open port list for FTP_server host (there is one open port – 21), since in accordance to port forwarding table incoming connections to IP 195.19.200.2:21 (where 21 is destination port) are forwarded into 192.168.0.11:21. Thus malefactor determines availability of one open port and he can attack it with "SYN flood" assault action. After performing second reconnaissance script ("Nmap OS"), malefactor receives information that does not allow to perform any assault action. After performing third reconnaissance script, malefactor can use three assault actions: (1) password searching ("FTP dict"); (2) denial of service attack ("ServU-MKD"); (3) privileges escalating ("ServU-MDTM"). First two actions are final. Third action allows malefactor to get administrator privileges and all FTP_server host information. Administrator privileges allows malefactor to go into the host and to attack other hosts.

Malefactor finds out that real FTP_server host IP (192.168.0.11) does not coincide with 195.19.200.2. Therefore, there is port forwarding in the network, and malefactor is at other subnetwork. This fact is critical to malefactor when he decides to change his location to the captured FTP_server host. Malefactor changes location and performs "Ping Hosts" action. He finds out that there are four hosts and consequently analyzes them with above-mentioned scheme. In addition he can get administrator privileges at hosts Firewall_1 and Firewall_2 because they trust to FTP_server.

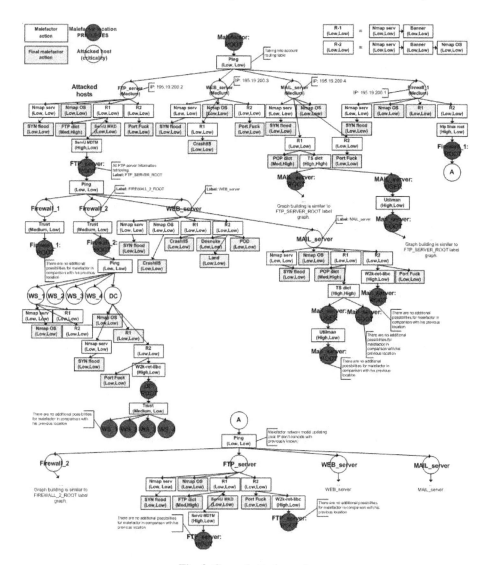

Fig. 3. General attack graph

Some of the security analysis results are as follows: Network bottlenecks – Firewall_1, FTP_server, ... ; Critical vulnerabilities – NTP_LINUX_ROOT, Serv-U MDTM, ... ; Graph has routes and threats with high mortality (for example, route Malefactor-Ping-FTP_server(Nobody)-Nmap serv-Banner-ServU MDTM-FTP_server(Root) ...); *SecurityLevel=Red*. The computer network security level does not satisfy user's requirements (better than Orange) and requires immediate actions for eliminating of the revealed software vulnerabilities and security policy bottlenecks.

8 Conclusion

The paper offered the approach and software tool for vulnerability analysis and security level assessment of computer networks, intended for implementation at various stages of a life cycle of computer networks. Offered approach is based on construction of attack graphs and computation of different security metrics.

The suggested approach possesses the following *peculiarities*:

- Usage for security level evaluation of integrated family of different models based on expert knowledge, including malefactor's models, multilevel models of attack scenarios, building common attack graph, security metrics computation and security level evaluation;
- Taking into account diversity of malefactor's positions, intentions and experience levels;
- Usage (during construction of common attack graph) not only of the parameters of computer network configuration, but the rules of security policy used; possibility of estimating the influence of different configuration and policy data on the security level value;
- Taking into account not only attack actions (which use vulnerabilities), but the common actions of legitimate users and reconnaissance actions which can be realized by malefactor when he gains certain privileges on compromised hosts;
- Possibility of investigating various threats for different network resources;
- Possibility of detection of "weak places" (for example, the hosts responsible for a lot of attack routes and the highest quantity of vulnerabilities);
- Possibility of querying the system in the "what-if" way, for example, how the general security level will change if the certain parameter of network configuration or security policy is changed or information about new vulnerability is added;
- Usage for attack graph construction of updated vulnerabilities databases (the Open Source Vulnerability Database (OSVDB) [11] is used);
- The "CVSS. Common Vulnerability Scoring System" [3] approach is used for computation of a part of primary security metrics.

The future research will be devoted to improving the models of computer attacks, the algorithms of attack graph generation and security level evaluation differing by accuracy and complexity, and experimental assessment of offered approach.

Acknowledgments. The research is supported by grant of Russian Foundation of Basic Research (№ 04-01-00167), Department for Informational Technologies and Computation Systems of the Russian Academy of Sciences (contract №3.2/03), Russian Science Support Foundation and by the EC as part of the POSITIF project (contract IST-2002-002314). Authors would like to thank the reviewers for their valuable comments to improve the quality of the paper.

References

1. Alberts, C., Dorofee, A.: Managing Information Security Risks: The OCTAVE Approach. Addison Wesley (2002)
2. Chapman, C., Ward, S.: Project Risk Management: processes, techniques and insights. Chichester, John Wiley (2003)
3. CVSS. Common Vulnerability Scoring System. URL: http://www.first.org/cvss/
4. Dantu, R., Loper, K., Kolan P.: Risk Management using Behavior based Attack Graphs. International Conference on Information Technology: Coding and Computing (2004)
5. Hariri, S., Qu, G., Dharmagadda, T., Ramkishore, M., Raghavendra, C. S.: Impact Analysis of Faults and Attacks in Large-Scale Networks. IEEE Security&Privacy, September/October (2003)
6. Jha, S., Sheyner, O., Wing, J.: Minimization and reliability analysis of attack graphs. Technical Report CMU-CS-02-109, Carnegie Mellon University (2002)
7. Lye, K., Wing, J.: Game Strategies in Network Security. International Journal of Information Security, February (2005)
8. Noel, S., Jajodia, S.: Understanding complex network attack graphs through clustered adjacency matrices. Proc. 21st Annual Computer Security Conference (ACSAC) (2005)
9. Netfilter/iptables documentation. URL: http://www.netfilter.org/documentation/
10. Ning, P., Cui, Y., Reeves, D, Xu, D.: Tools and Techniques for Analyzing Intrusion Alerts. ACM Transactions on Information and System Security, Vol. 7, No. 2 (2004)
11. OSVDB: The Open Source Vulnerability Database. URL: http://www.osvdb.org/
12. Ou, X., Govindavajhala, S., Appel, A.W. : MulVAL: A Logic-based Network Security Analyzer. 14th Usenix Security Symposium (2005)
13. Rieke, R.: Tool based formal Modelling, Analysis and Visualisation of Enterprise Network Vulnerabilities utilising Attack Graph Exploration. EICAR 2004 (2004)
14. Ritchey, R. W., Ammann, P.: Using model checking to analyze network vulnerabilities. Proceedings of IEEE Computer Society Symposium on Security and Privacy (2000)
15. Rothmaier, G., Krumm, H.: A Framework Based Approach for Formal Modeling and Analysis of Multi-level Attacks in Computer Networks. LNCS, Vol.3731 (2005)
16. Sheyner, O., Haines, J., Jha, S., Lippmann, R., Wing, J.M.: Automated generation and analysis of attack graphs. Proc. of the IEEE Symposium on Security and Privacy (2002)
17. Singh, S., Lyons, J., Nicol, D.M.: Fast Model-based Penetration Testing. Proceedings of the 2004 Winter Simulation Conference (2004)
18. Swiler, L., Phillips, C., Ellis, D., Chakerian, S.: Computer-attack graph generation tool. DISCEX '01 (2001)

Information Modeling for Automated Risk Analysis

Howard Chivers

Department of Information Systems, Cranfield University,
Defence Academy of the United Kingdom, Shrivenham, Swindon, SN6 8LA, UK
hrchivers@iee.org

Abstract. Systematic security risk analysis requires an information model which integrates the system design, the security environment (the attackers, security goals etc) and proposed security requirements. Such a model must be scalable to accommodate large systems, and support the efficient discovery of threat paths and the production of risk-based metrics; the modeling approach must balance complexity, scalability and expressiveness. This paper describes such a model; novel features include combining formal information modeling with informal requirements traceability to support the specification of security requirements on incompletely specified services, and the typing of information flow to quantify path exploitability and model communications security.

Keywords: security, risk, model, information, threat, service-oriented, communication.

1 Introduction

Security Risk analysis provides a criterion for the value of security in the business and social context of a system. It is the only viable cost benefit justification for security controls, and is the established basis for information security management standards [1] and methods [2]. There are a range of problems in applying systematic risk analysis to large distributed service-oriented systems, one of which is developing an analytic model which is able to support automated threat analysis.

The SEDAN (Security Design Analysis) framework has been developed to support the risk analysis and security design of large distributed systems. At the core of the framework is an information model, which integrates a system design, usually expressed in UML, with a specification of the security environment, including attackers and security objectives. The primary function of this model is to support automated threat path analysis - finding paths from attackers to critical assets - which is at the heart of risk analysis.

The design of this information model is a compromise between the need for efficiency and scalability, and the need to accurately model a diverse range of security objectives and requirements. The information model described here efficiently interprets the information flow in a system as a graph; however, the needs of risk analysis and requirement modeling have resulted in novel features in how the graph is constructed and used. These include combining a generic model of information-flow with informal requirements traceability, allowing the specification of security requirements on incompletely specified sub-systems, or services, and typing of

H. Leitold and E. Markatos (Eds.): CMS 2006, LNCS 4237, pp. 228–239, 2006.

information within the model, to distinguish threat path exploitability and allow the specification of communications security requirements.

The contribution of this paper is that it describes an approach to information modeling specifically designed to support risk analysis. In order to ensure scalability and allow the specification of a wide range of security requirements the model has a number of novel features, including combining formal modeling with informal requirements traceability, and the typing of information flow to distinguish path exploitability, and model communications security.

The information model described in this paper has already been applied in practice, by the production of supporting tools, and their use in the analysis of a complex industrial distributed system [3]. For reasons of space only the information model is described here; the specification of security requirements in the SEDAN framework is published separately [4], together with a worked example. A detailed account of the framework and its application is also available [5], which includes a formal account of the model described in this paper.

This paper is organized as follows: Following a brief description of related work, section 3 describes the information model, how it is related to a system design, and the motivation for combining formal information modeling with informal requirements traceability. Section 4 describes an important extension to the basic model: information typing. Section 5 discusses possible limitations in graph-based modeling and how they are overcome, and section 6 concludes the paper.

Definitions

In this paper a security objective or protection objective is an unwanted outcome for a particular asset, also known as an asset concern; such objectives are the goals of threat path analysis. A security requirement, or control requirement is an operationalized requirement (e.g. an access control), that is part of the specification for a system component, usually in the form of a functional constraint.

This paper is concerned with system protection; security goals also require functions, such as intrusion detection, but these are beyond the scope of the paper.

2   Related Work

Risk analysis approaches were reviewed by Baskerville [6] over a decade ago, and his analysis is still relevant today. He characterizes current methods (including tools, such as CRAMM [7]) as engineering based, but failing to integrate risk analysis with the engineering models of the systems they support. He identifies the need to use abstract models of systems, which allow a combination of security design and stakeholder analysis, independent of physical implementation. This characterization of the problem is one of the motivations for the SEDAN framework, since it identifies the scope for abstract modeling to make a fundamental contribution to risk analysis.

Recent work on risk is typified by the European CORAS research project [8]; this has sought to integrate risk and engineering models by providing process metamodels and threat stereotypes for UML, and by investigating how various methods from the safety and risk community (e.g. failure mode analysis, fault tree analysis) can be utilized in e-commerce risk assessment. Essentially it provides a documentation base for risk analysis, but no new modeling, and while there are many proponents of threat modeling or analysis (e.g. [9]), these describe good practice and principles, but not abstract models that allow systematic tool support.

Related work, such as UMLSec [10] builds on the formal semantics of UML to allow the complete specification of security problems, which can then be subject to formal proof, or exported to a model-checker. This work is typical of a wide range of formal approaches to security; it is promising for small hard security problems, such as protocol analysis, but there is little evidence that this approach will scale to large-scale practical systems, or that it can accommodate risk as a criterion.

Security risk management is essentially a form of requirements engineering, and the goal-refinement community are active in developing new requirements management models [11], some of which are tool supported. However, this work has yet to accommodate risk metrics, or threat analysis.

In summary, the creation of an effective analytic model for risk analysis is an important open question; the approach described in this paper is unique, since it systematically combines formal modeling and informal requirements traceability.

3   Modeling the Information in a System Design

The purpose of the information model described in this paper is to enable the systematic, automated, discovery of threat paths, and the calculation of other risk-based metrics, in a high-level service-oriented system design. The starting point for the information model is therefore what is represented in such a design:

- the structure of the system: its services (or sub-systems) and data structures;
- interfaces to these services, including the messages that they support; and,
- communication between services: which services are invoked by others.

The following sections describe the basic information model, including how it is mapped to a system, the need to represent security requirements, and resolving the problem of incompletely specified system behavior by combining a generic model of service behavior with traceability to informal security objectives.

3.1   Mapping the System to an Information Flow Graph

Threat path discovery is a form of model checking: it is necessary to expand the information model into a graph, determine paths that correspond to threats, and relate these results back to the system. Following paths in a graph also corresponds to an intuitive model of threat analysis, so the information model is formulated as a graph which represents the information flow in the system.

The system is divided into information carriers and behaviors, which are mapped to graph vertices and directed edges, respectively. Information carriers are data, messages, or events in the functional model; behaviors include system functions or services. The graph is directed, and information paths in the graph may be cyclic. Users (strictly, user roles) are modeled as sources or sinks of information. For example, consider the simple system presented in the UML model in fig 1.

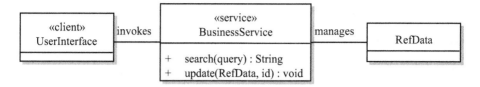

Fig. 1. An Illustrative System

In fig 1 there are two services, the first, stereotyped <<*client*>> is directly accessible by users, and the second (*BusinessService*) provides operations to search and update business data assets (*RefData*). In this system the information carriers are the messages between the two services, that invoke or return data from their operations, the data asset, and messages that flow directly to users.

The corresponding information graph is shown in fig 2, in which the services (*s1, s2*) encapsulate graph edges (internal to the services and not shown), and the vertices (*va ... vg*) represent information carriers that are identifiable in the system. The figure is annotated to show how the graph is related to the system in fig 1.

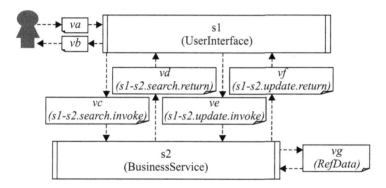

Fig. 2. The Information Graph: vertices represent information carriers such as messages or events; service behavior is characterized by edges that carry information between vertices

Fig. 2 illustrates the mapping between a system and the information model:

- vertices represent information carriers, and have exactly one service as their input and one service as their output;
- the edges of the graph are partitioned between the services of the system. The only information flow between services is via vertices; and
- users are represented as active subjects, similar to services, in that they can invoke operations in client services. However, unlike services, they are sources or sinks of information, rather than graph edges.

Graph vertices are derived directly from information carriers identified in the system model. For example, *vc* and *vd* model the call and return of the *search* operation in the *BusinessService* service (*s2*), invoked by the *UserInterface* service (*s1*). Note that from an information perspective there is usually no need to distinguish the fine grain data structure within a message in information terms (e.g. multiple parameters in the *update()*), but there is a need to represent events that carry no data, such as the *void* return from *update()*; this is discussed further in section 4. This mapping does not imply that messages to and from services are necessarily synchronous.

Vertices can equally represent information exchanged with system users (*va, vb*), or data that is part of a service's state (*vf*); from a threat path discovery perspective, all assets of concern are mapped to graph vertices.

The edges of the graph capture system behavior or functionality, but as noted above, the behavior of services in a system design may be unspecified. Unless a service is constrained by a security requirement, it potentially routes information from all its inputs to all its outputs. For example, in service *s2* it is possible to distinguish nine distinct behaviors (and hence, graph edges) that represent information flow between {*vc, ve, vg*} and {*vd, vf, vg*}. In the absence of security requirements that restrict this behavior, this generic model of a service is used to define the graph edges.

A consequence of this strict division between information carriers and behavior is that a service is never mapped to a graph vertex, which suggests that it can never be the target of a threat path. This is discussed further in section 5.3.

3.2 Modeling Security Requirements

This section describes how the information model supports security requirements; a more detailed explanation of how requirements are specified is published separately [4], together with a worked example.

An essential part of risk management is evaluating a proposed protection profile (a set of security, or control, requirements) to identify residual threats. Some security requirements can be represented in terms of a system design, and some are more difficult. For example, access controls are, in principle, straightforward to specify and model, since they constrain messages that can be identified in the system. However, the specification of constraints on the behavior of services is not as straightforward.

For example, consider the system information graph shown in fig 3; the same symbols are used to denote services and data as in fig 2, but vertices that are not important for the discussion are unlabeled, solid arrowed lines show invocation, dashed lines show other relevant information flow.

In fig. 3, an attacker has access to two services (*sa, sb*) that invoke further services (*sc,sd,se*) to update a data asset (*va*).   The security objective is to prevent unauthorized modification of the data asset. There are several paths between the attacker and the asset, so there are a number of options for how the resulting threats can be defended. It is obvious by inspection that unless more is known about the behavior of the services, a single security control is not sufficient to protect all the possible threat paths; at least two are needed, for example *ra* and *re*.

These two control requirements are different in type. For example, service *sa* may be a management interface which normal users (including this attacker) do not need to access: *ra* is an access control. Requirement *re* is unlikely to be as straightforward, since given the system configuration, it is unlikely that all accesses between *sb* and *se* can be prohibited. The requirement *re* must constrain the operation invoked by *sb*, rather than prevent it; perhaps by allowing read-only access to *va*.

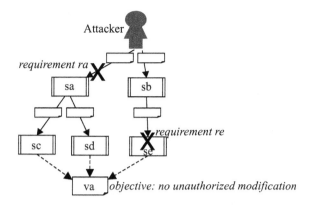

Fig. 3. Attacking a Data Asset via Services

This example highlights the difference between requirements that constrain specific elements of the system model (e.g. access controls, to restrict possible messages) and constraints on elements of the system that are not fully specified[1]: the behavior of services. This also underlines the difference between security and functional requirements: it may have been the intention of the designer that service *se* provides read-only access to *va*, but a systematic analysis will ensure that this is identified and documented as an essential security requirement.

One approach to specifying control requirements on service behavior is to first fully specify the behavior of each service, but this anticipates the design process, reduces the options available to an implementer, and may suffer from the scalability difficulties associated with the use of formal methods in large systems [12]. This paper describes an alternative: the use of generic information-flow constraints, complimented by traceability to informal security objectives. This is the approach described in the next section.

3.3 Deferred Requirements

In fig. 3, requirement *re* has three main components:

service *se* ... must protect *va* from *sb* ... to prevent unauthorized modification

The function of service *se* is not specified, so this requirement cannot be formalized within the domain of discourse provided by the system design. A generalized information flow constraint captures the first two parts of the requirement (constrain information flow between *sb* and *va* in service *se*); however, the security objective (prevent modification) is an important clarification of the requirement.

This type of requirement is described as deferred, because the semantics of the protection objective are informal, and can be properly interpreted only in terms of a detailed functional design or implementation. Deferred requirements have three parts:

---

[1] In a system design the *interface* to a service (i.e. the messages it receives) is specified, but this is distinct from its *behavior* (i.e. what it actually does).

- the service which must implement the requirement;
- the information context: which messages at the interface to the service, or which assets within the service's state, are constrained; and
- informal semantics, which are specified by reference to the security objective for the associated asset[2].

This approach is a compromise between a formal model of security, and informal requirements management. The former requires a system to be modeled in sufficient detail to allow the specification of any functional constraints, the latter does not benefit from a fully systematic analysis. Essentially, the formal information model defines where requirements are placed in the system, but traceability to informal security objectives are used to clarify what the requirements must achieve.

3.4   Graph Sets

The semantics of deferred requirements are not fully defined in the information model, since part of their specification is informal. As a result, deferred requirements associated with different security objectives are not necessarily comparable. In other words, an information flow constraint traceable to one security objective does not necessarily protect another. For this reason the SEDAN information model is not a single information graph, but a set of graphs, one for each security objective.

This feature is a technical issue for the implementation of the associated model-checker, but does not essentially change the underlying efficiency of the graph-based modeling approach.

3.5   Diverse Security Objectives

The combination of information modeling and informal semantics is able to accommodate a wide range of security objectives. Confidentiality can be interpreted directly in information-flow terms, but most other security objectives are not as easily expressed; for example, integrity has a wide range of different interpretations [13], including no unauthorized changes, or maintaining provenance, or consistency.

These different types of integrity can be treated in a uniform way: the information model allows the discovery of threat paths from attackers to related assets, and security requirements can be placed on these threat paths to protect the security objective; this resolves the problem in information terms, but does not distinguish between different integrity objectives. The implementer is able to determine the detailed protection requirement by traceability to the informal security objective.

This pattern of formal path discovery, and informal requirements traceability, therefore supports a wide range of different security objectives. (See also section 5.2)

4   Information Typing and Communications Security

One security requirement that could be used to preserve integrity (see previous section) is authentication: the source of data is accredited, preventing an attacker from

---

[2] This does not imply that each asset has distinct security objective; security objectives may apply to groups of assets, or be traceable to higher level goals that specify their purpose.

injecting or substituting false data. However, if the objective were availability, then data authenticity would not protect against a denial of service attack involving high volumes of invalid data.

This is an example of a threat which can be transmitted via system events, or traffic flow; in risk analysis it is necessary to distinguish traffic flow from information carried by data, since they support different threats. Vertices in the information model are therefore typed to characterize the threat paths that may be supported by the vertex. The base types are *data* or *void*[3], corresponding to information carriers that support data or traffic, respectively.

However, the value of information typing extends beyond the need to characterize threat paths; it also allows communications security requirements to be represented in the information model, and this is described in the remainder of this section.

Modeling Communications Security

Attacks via implementation mechanisms (e.g. buffer overflow, or direct access) are common and important, and one purpose of communications security[4] is to protect against such attacks. The associated threat paths can be determined by evaluating the impact of an attacker with direct access to a graph vertex. This requires the type of a vertex to be further qualified by its accessibility to an external attacker. Three additional types are needed: *Confidentiality*, *Integrity*, and *Virtual*. Fig. 4 shows how they are used to model confidentiality; data flow is shown dashed, and traffic flow dotted.

In fig.4, vertex *Vx* represents an encrypted message, resulting in traffic flow from service *s1* to *Vx* (1). An attacker with direct access to the message, for example by wiretapping, is able to extract only traffic information (2). However, if the attacker injects data (3) then this may be inadvertently accepted by service *s2* (4). A *Confidentiality* vertex therefore supports a traffic flow from its source to an external attacker, and a data flow from the attacker to the destination service. The reverse pattern applies to an *Integrity* vertex: an external attacker is able to obtain data, but not misrepresent it.

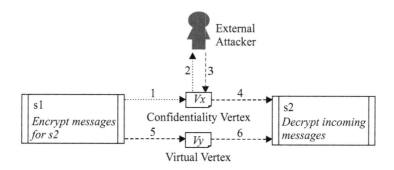

Fig. 4. Vertex Types used to Model Confidentiality

---

[3] *Void* is named because many related system events are messages with void parameter sets. For example, vertex *vf* in fig 2.

[4] Communications security is also used to protect end-to-end messages from intermediate services; this is also accommodated by the model described here.

The *Confidentiality* type characterizes the communication level message from *s1* to *s2*; however, *s2* obtains data from this message by decryption, unlike an external attacker or intermediate service (not shown), so it is also necessary to add a direct auxiliary information path (5-6 via *Vy*) to model the recovery of message content. This vertex has a special type, *Virtual*, which corresponds to the message layer in the system model, conveys data, and is inaccessible by an external observer.

This describes the main features of information typing: it allows the information model to distinguish between traffic and data flow in the system, and hence allows the modeling of communications and message security in an information graph.

5   Potential Limitations

The interpretation of a system as a set of information graphs provides an efficient basis for threat path analysis on large systems; however, it makes some explicit assumptions about the implementation, and is less expressive that other possible formal expressions of information flow (e.g. as a set of traces). The following sections review the key assumptions and possible limitations.

5.1   Critical Modeling Assumptions

The structure of the information model embodies critical assumptions that must be maintained in an implementation if the risk analysis is to remain valid; these are that:

- a system can be decomposed into services;
- the only information flow between services is via identified information carriers; and
- the structure of services and information carriers is consistent between the system design and its implementation.

These assumptions are appropriate to service-oriented systems that comprise services communicating via messages; individual services may be deployed to physically separate servers, but however they are deployed the implementation is likely to preserve the isolation between services. However, if deployed services are able to communicate via a mechanism that is not present in the design, then this introduces a behavior that was not anticipated in the analysis, with undefined consequences for security.

This is a special case of the general principle that information flow in an implementation must be a subset of that analyzed [14]. These implementation constraints are therefore normal for security analysis and modeling generally, and the system structure described here is well adapted to service-oriented systems.

5.2   Information Representation Limitations

The information model is a graph which is mapped to information carriers in the system design. Such a graph can be made as expressive as necessary, by expanding the number of vertices to enumerate properties of interest; however, in practice there is a need to balance scalability and expressiveness, so the properties enumerated are limited: vertices are distinguished by source, destination or ownership. As a consequence the information graph does not directly model sequence or time, and this potentially limits the security objectives that can be analyzed.

Consider such an objective: two different users are required to perform an action in sequence; one originates a purchase order, and the second must approve it. The security objective is to avoid an incorrect transaction sequence, because the organization wishes to prevent an approver writing 'blank cheques'.

One or more security requirements are needed in the paths between these users and the payment. The information model can be used to show that a security requirement is correctly placed, because threat paths can be identified between the users and the payment; the system implementer is able to determine that this requires 'correct sequence', because the requirement is traceable to the informal security objective.

The essence of this example is that the information model is able to solve the critical problem in systematic design – the placement of security requirements – without necessarily needing to fully interpret the requirement. This approach is just as effective for temporal constraints as for other constraints on behavior, and mitigates the absence of formal temporal modeling.

5.3 The Representation of Services

A service is modeled as pure behavior: a set graph edges that convey information flow (see section 3.1). It is the vertices of the graph that represent identifiable data items in the system, and are the potential targets of attack. This strict division between information carriers and behavior seems to suggest that a service can never be the target of a threat path.

This potential issue can simply be avoided by explicitly modeling a service's state as data, and identifying that data as a security-critical asset. This approach is occasionally necessary in practice; for example, if the algorithm or software used to perform a service is itself confidential. However, experience suggests that security objectives that system stakeholders wish to assign to services are often misplaced. The most common example is availability: intuitively 'availability of a service' is an appealing security objective; however, what is usually required is availability of the results of the service to the user. From a modeling perspective it is preferable to identify the result as the asset of concern, since the whole information path, including the service, is then the subject of analysis. An abstract model should clarify important aspects of the target system, and this strict mapping of services to behavior prompts the user to identify unambiguously the targets of protection.

6 Conclusions

This paper describes an information model that has been developed to support security risk analysis. The essential structure of the model is an information graph, in which vertices correspond to identifiable information carriers in the system (e.g. messages) and edges represent service behavior. This approach supports a direct mapping between the system design and the information model, and allows the efficient and intuitive analysis of threat paths.

Compared to fully formal system modeling, this approach offers considerable scalability and efficiency, but is potentially less expressive; this problem is overcome by two novel features: the use of informal objectives to clarify security requirements, and information typing.

The need for informal semantics arises because the behavior of services is not defined in a high-level system model. Security requirements on the behavior of such services are specified by a combination of a generic information-flow constraints, and are traceable to the security objective which they support. In effect, the information model determines threat paths and specifies the position of security requirements, while the specific form of protection is clarified by the informal security objective.

This mixture of formal and informal requirements management accommodates a wide range of different security objectives, and also has technical consequences: the information model is a set of graphs, one for each security objective.

Information typing distinguishes between data and traffic flow in the system; this characterizes the exploitability of different threat paths, and allows the modeling of communications, and message-based security.

The information model described in this paper has been used to support practical risk-analysis tooling, and the analysis of a substantial industrial system [3]. The remaining open questions are not concerned with the underlying model, as described here, but with suitable models for established security requirements (similar to fig 3), which are often patterns within the information model.

Acknowledgements

This work was carried out as part of a Royal Academy of Engineering Senior Research Fellowship. We are also grateful to the anonymous referees for their perceptive and constructive contribution to this paper.

References

1. Information Security Management Part 2 Specification for information security management systems, British Standards Institution, BS 7799-2:1999.
2. Risk Management Guide for Information Technology Systems, National Institute of Standards and Technology (NIST), SP 800-30. January 2002. http://csrc.nist.gov/publications/nistpubs/800-30/sp800-30.pdf (accessed January 2006)
3. Chivers, H. and Fletcher, M., Applying Security Design Analysis to a Service Based System. Software Practice and Experience: Special Issue on Grid Security, 2005. **35**(9). 873-897.
4. Chivers, H. and Jacob, J. Specifying Information-Flow Controls, Proceedings of the Second International Workshop on Security in Distributed Computing Systems (SDCS) (ICDCSW'05), Columbus, Ohio, USA. IEEE Computer Society, 2005; 114-120.
5. Chivers, H., Security Design Analysis, Thesis at Department of Computer Science, The University of York, York, UK, available on-line at http://www.cs.york.ac.uk/ftpdir/reports/YCST-2006-06.pdf, (accessed July 2006). p. 484. 2006
6. Baskerville, R., Information Systems Security Design Methods: Implications for Information Systems Development. ACM Computing Surveys, 1993. **25**(4). 375-414.

7. CRAMM Risk Assessment Tool Overview, Insight Consulting Limited, available at http://www.cramm.com/riskassesment.htm (accessed May 2005)
8. Dimitrakos, T., Raptis, D., Ritchie, B., and Stølen, K. Model-Based Security Risk Analysis for Web Applications: The CORAS approach, Proceedings of the EuroWeb 2002, St Anne's College, Oxford, UK. (Electronic Workshops in Computing). British Computer Society, available on-line at http://ewic.bcs.org/conferences/2002/euroweb/index.htm (accessed January 2006), 2002.
9. Swiderski, F. and Snyder, W., Threat Modelling. Microsoft Professional. 2004: Microsoft Press.
10. Jürjens, J. Towards Development of Secure Systems Using UMLsec, Proceedings of the Fundamental Approaches to Software Engineering : 4th International Conference, FASE 2001 : Held as Part of the Joint European Conferences on Theory and Practice of Software, ETAPS 2001, Genova, Italy. (Lecture Notes in Computer Science vol 2029). Springer-Verlag, 2001.
11. Kalloniatis, C. Security Requirements Engineering for e-Government Applications: Analysis of Current Frameworks, Proceedings of the Electronic Government: Third International Conference, EGOV 2004, Zaragoza, Spain. (Lecture Notes in Computer Science vol 3183 / 2004). Springer-Verlag, 2004; 66-71.
12. Schaefer, M. Symbol Security Condition Considered Harmful, Proceedings of the IEEE Symposium on Security and Privacy, Oakland, CA. IEEE Computer Society, 1989; 20-46.
13. Mayfield, T., Roskos, J. E., Welke, S. R., and Boone, J. M., Integrity in Automated Information Systems, National Computer Security Center (NCSC), Technical Report 79-91. http://www.radium.ncsc.mil/tpep/library/rainbow/C-TR-79-91.txt (accessed January 2006)
14. Jacob, J. L. On The Derivation of Secure Components, Proceedings of the 1989 IEEE Symposium on Security and Privacy. IEEE Computer Society, 1989; 242-247.

Towards Practical Attacker Classification for Risk Analysis in Anonymous Communication

Andriy Panchenko and Lexi Pimenidis*

RWTH Aachen University,
Computer Science Department - Informatik IV,
Ahornstr. 55, D-52074 Aachen, Germany
{panchenko, lexi}@i4.informatik.rwth-aachen.de

Abstract. There are a number of attacker models in the area of anonymous communication. Most of them are either very simplified or pretty abstract - therefore difficult to generalize or even identify in real networks. While some papers distinct different attacker types, the usual approach is to present an anonymization technique and then to develop an attacker model for it in order to identify properties of the technique. Often such a model is abstract, unsystematic and it is not trivial to identify the exact threats for the end-user of the implemented system. This work follows another approach: we propose a classification of attacker types for the risk analysis and attacker modelling in anonymous communication independently of the concrete technique. The classes are designed in the way, that their meaning can be easily communicated to the end-users and management level. We claim that the use of this classification can lead to a more solid understanding of security provided by anonymizing networks, and therewith improve their development.

Finally, we will classify some well known techniques and security issues according to the proposal and thus show the practical relevance and applicability of the proposed classification.

Keywords: anonymous communication, attacker model, risk analysis.

1 Introduction

The primary goal in anonymity networks is to achieve sender anonymity, recipient anonymity, or both. The term *anonymity* is often defined as "the state of not being identifiable within a set of subjects, the anonymity set" [23]. This definition implicitly assumes a system state where there is either no attacker or the attacker is not successful. The task to estimate whether an attacker will be successful in breaking a real system or not is done as a part of the security evaluation or risk analysis. The most critical part of this is to properly define a realistic attacker model. If the chosen attacker model is too powerful - most of the protection techniques will necessarily fail, if the attacker model is too weak - the system will inevitably provide false and undesired means about protection level of its users.

---

* The authors are funded by the European Commission's 6th Framework Program.

H. Leitold and E. Markatos (Eds.): CMS 2006, LNCS 4237, pp. 240–251, 2006.

Especially in the field of anonymous communication there exist a large number of attacker models. Most of these are describing the actual capabilities of the attacker, not considering the power needed in real life to achieve the proposed capabilities. A common example is the passive global observer. We agree that this model is needed and interesting for mathematical analysis, however end-users should be aware that theoretical results based on this analysis are not representative in real scenarios: an attacker having the capabilities to intercept traffic at the global scale can typically also easily alter and manipulate the traffic and, therewith invalidate the results of the analysis and protection vision of the end-user. From another perspective, it is not realistic for an average end-user to defend against an adversary that is capable of observing the whole worldwide network, because such a powerful adversary can make use of more efficient means in order to obtain the same information.

Only few systems for anonymous communication can be proven to be secure against very powerful attackers, given that the implementation is not faulty. A good example is a DC-network[4] which is known for its high security level. On the other hand, there are systems that provide security against weak attackers but fail against the strong ones. We call the resulting state *practical anonymity* (with regard to the thwarted attackers).

Most of the existing attacker models arised in the way, that at first an anonymization technique was presented and then the model was suggested in order to identify properties of the system. This often resulted in an unsystematic and abstract outcome of the attacker representation. We thus propose a new method for attacker characterization that is less abstract and more practical, therefore can be easily communicated to the end-users. The classification shall also provide a proposal for a simplistic measure of *quality of protection* in anonymous networks. In this specific work we will develop an attacker classification for anonymous communication systems and show an example of its application. At this point we want to clearly state, that the proposed classification is not strict: it is possible to classify in a different way. The same applies to the number of categories and the attacker classes they describe. Herewith we want to give an incentive to the community in the area of anonymous communications to think about realistic attacker models and link them to existing attacker descriptions rather than to replace existing classifications. This work is thus an *overview* on attacker models, their *classification* and *applicability* to current implementations.

1.1   Contribution

While it is theoretically feasible to defend against a nearly arbitrarily powerful attacker, it seems to us that such a system would be so slow and prone to denial of service attacks that the amount of users willing to use it would be very small. On the other hand, anonymizing networks are strongly in need for a large number of users to increase the size of anonymity set. Thus, it is not a good choice to defend against arbitrary powerful attackers. Therefore our work's aim is to allow the users to identify the attacker types they want to protect themselves from (*practical anonymity*). Having identified them, it is possible to look for techniques that would provide the desired degree of protection.

Our contribution to this topic is twofold:

1. We propose a classification for categorizing attacker types.
2. We show the applicability of the model with a short analysis of the strength of anonymizing techniques as well as some widely known attacks on them.

2   Related Works

To the best of our knowledge there is no paper dedicated explicitly to the attacker classification for anonymous communication, although all major papers in this domain define one or more attacker models. In this section we will give an overview of existing attacker models. Please note also that the majority of these papers primarily proposed a technique for anonymization and developed attacker models in order to distinguish their work from previous results (i.e. in order to identify properties of the new system). Thus these models are quite unspecific with regards to real systems.

In general it is assumed in literature on traffic analysis and anonymous communication that the attacker knows the infrastructure and strategies that are deployed[1]. This assumption is similar to those made in cryptology, where it's commonly assumed that an adversary knows the algorithms that are used.

Some attacker models in literature are quite simple. While this can be correct from a theoretical point of view, it arises difficulties in case of the risk estimation in the real world settings. In [29] the adversary is described as a participant that collects data from its interactions with other participants in the protocol and may share its data with other adversaries. [26] describes an attacker as some entity that does passive traffic analysis and receives the data by any mean that is available. These kinds of attacker models might be interesting in certain special cases but are difficult to generalize and identify in a real system: depending on the influences these attackers might have - they can be completely different entities. So, for example, they can be a secret service or standalone hacker, each being a different threat to the end-user. And the means that should be taken in order to provide the protection depend on the concrete threat entity.

A more general attacker categorization is given e.g. in [16]. Authors introduce three classes of attackers with increasing amount of power and capabilities, namely the *global external passive attacker*, the *passive attacker with sending capabilities* and the *active internal attacker*. While this distinction makes sense in the context of the paper [16] because it helps to show a difference between Mixmaster and Stop-and-Go-Mixes, the difference is marginal to virtually non-existing in real systems. We agree that a purely passive attacker is different from an attacker that also participates in the network and is possibly detectable. On the other hand, it's quite unlikely that an attacker that has global access to

---

[1] Since this is a commonly used assumption we intentionally omit a long list of references. See for example http://www.freehaven.net/anonbib/

network lines does *not* also have the possibility to inject messages. So, the first two attacker types wouldn't differ in their capabilities in real systems but rather in the decision whether to make use of all their features.

The same applies to [25], where the authors propose to split a *global active attacker* into the one that can only insert messages, and the one who can delay messages. If an attacker is able to deterministically delay messages in a real system, he will also be able to insert messages. On the other hand, if an attacker is able to insert messages in a system and observe their effect, he is most probably present on some of the system's lines and thus able to delay messages.

A more detailed list of adversaries can be found in [13], where four attacker types are listed: the *eavesdropper*, the *global eavesdropper*, a *passive adversary* and an *active adversary*. Again there will be little difference between e.g. the global eavesdropper and an global adversary in practice.

The most systematic listing of attacker types for theoretic modelling is found in [24], where Raymond introduces three dimensions of attackers:

internal-external Attackers can be distinguished on whether they are participants in the network or not.
passive-active Attackers can actively change the status of the network or remain passive.
static-adaptive Attackers can't change their resources once the attack has started or they can continue to build up their capabilities.

An additional dimension is given by Pfitzmann in [22]: active attackers can either limit their actions, follow the protocol and thus reduce the chance of being detected, or trade-off their stealth in favor to more powerful attacks by committing actions that are not part of a network's protocol.

The most realistic attacker model can be found in [28] where not only the method of attack is provided (ranging from an observer to a hostile user or a compromised network node) but also the extend of the attacker's influence on the network (i.e. whether it's a single node or some large parts of the network).

There is a large body of survey and classification material associated with risk analysis e.g. in [14,1]. However, most of them define a set of skills, resources, etc. of an attacker, without binding these to real entities and not focusing on the practical representation.

3 Attacker Classification

The central idea of the proposed classification is to give an overview of possible common attackers in *real networks* and classify their strengths, weaknesses and capabilities. It is designated to help management level and end users to do their own personal risk analysis. A reason for this is that it is in general not an adequate choice to defend against the most powerful attacker that is possible. This is especially the issue in the area of anonymous communication where every added piece of protection reduces usability. Our classification can be used for end-users and in business applications to properly communicate the threat of certain known attacks. We will evaluate this estimation and show an example of its application in Section 4.

To achieve a better understanding of the adversary faced with, we propose to classify the formerly abstract attacker types (e.g. passive/active attackers) in a new grid. We still assume that an attacker has the knowledge about the infrastructure of the network and its algorithms. This is reasonable because all major contemporary implementations of anonymizing networks are either open source, well documented or can be downloaded and reverse-engineered. We also assume that an attacker knows about all major attacks that have been discussed and published in the literature.

Every attacker is typically also able to conduct passive as well as active attacks. However, we can neither estimate nor model a potential attacker's skills that go beyond the current state of published attacks[2]. But we might consider attacker class conditioned bounds in order to estimate the amount of required resources for a successful attack depending on the information theoretical calculations ([18]).

The attributes that distinct most real life attackers are the *amount of computational power* and the *amount of influence that the attacker has on the network*. The latter correlates most often with the number of nodes and links that the attacker controls or which are within his reach. Furthermore, computational capabilities are not as relevant in today's scenarios because cryptography is usually too strong to be broken by NGOs[3] and computational breaking of other primitives is only seldom preliminary to attack an anonymizing system.

3.1 Proposed Classification

We hereby propose the following classification of attacker types. These are not chosen by the network's infrastructure or topology, but rather as entities and social stereotypes participating in, affected by or being interested in a transaction between two parties using an anonymizing network. However this should not be regarded as a restriction, since it is unlikely that these entities and social stereotypes will be replaced or become irrelevant in the future, even if the underlying networks change.

It is assumed as an unconditional requirement that the user's terminal is under his own control and cannot be compromised by any other party. Otherwise it is trivial breaking the user's privacy and anonymity.

0. External Party. The least powerful attacker has no control of any computer between the two communicating parties. While this kind of attackers are hardly worth being called so, there should be still taken measures to prevent them from gaining information.

---

[2] But we will consider it in the future work to keep the classification up-to-date.

[3] Non Governmental Organizations

Note that external parties can be very powerful, e.g. competitors in international trade, but unless further actions are taken to increase their influence on anonymizing networks, their influence is limited.

1. **Service Provider.** This class of attacker stands for the user's communication partner. In some scenarios it is desirable to omit the disclosure of the sender's true identity. This attacker is technically bound to the receiving end of the communication and its close neighborhood.

2. **Local administration.** This class of attackers can manipulate and read everything in the close network environment of the user[4]. These capabilities can be very powerful if the user blindly trusts all the transmitted and received data or does not care about protection. On the other hand, this attacker can be easily circumvented once the user is able to establish a secure connection to an outside trusted relay.

3. **ISP.** The next powerful attacker has access to the significant larger area of computers in the vicinity of the user. The amount maybe so large that it can even be a non-negligible part of the whole global network. It is thus possible that a major number of relays on the way to the communication partner is within the reach of this class of attacker.

4. **Government.** This adversary has the power to access not only a significant portion of all networks but also has large resources to fake services, break simpler encryption schemes[5] or prohibit access to specific services. This adversary might also take measures that violate existing laws to a certain extent and has the power to draw significant advantages from doing so.

5. **Secret Services.** are forming the highest class of an adversary. They can be assumed to either have access to most parts of the global networks or they can get the access if they think it's necessary for their operation. This class of attacker is also not bounded by any kind of laws. It should be mentioned that the latter two types of attackers will probably not refrain from using non-technical methods to get information - this includes but is not limited to the physical capture of nodes. It is noteworthy that some countries deploy their Secret Services for industrial espionage.

We deliberately don't specify the classes of attackers in more detail, but rather leave them as categories that are intuitively understood by researchers as well as by the end-users. Note that these classes must not be strict: seamless transition is allowed.

For example, traditional law enforcement can be seen as an attacker split up on classes 4 to 5. Furthermore, Figure 1 gives some techniques for anonymous communication and specifies the highest class of attacker they protect against.

---

[4] Think of sniffing data, manipulated DNS-responses, man-in-the-middle attacks on TLS-secured connections, denial of access to anonymizing networks to force plain communication, and much more.

[5] The German Federal Office for Information Security factored the RSA-640 number in September 2005 and single-DES is known to be weak for decades: http://www.rsasecurity.com/rsalabs/node.asp?id=2092

Technique	Defends against
Encrypted Communication	class 0 = External Party
Open Proxy Relay	class 1 = Service Provider
Encrypted Proxy Relay	class 2 = Local administration
JAP, Tor	depending on the configuration: class 2 to 3
Mixmaster[6]	class 3 to class 4

Fig. 1. Example: Techniques and the attacker types they defend against

From our point of view, the minimum requirement for an anonymizing network should be to defeat from attackers of class 0 upwards to the class 2 or 3. While it seems currently to be infeasible and to some people not desirable to protect all end-users from attackers of class 4 and higher ones, we list these for completeness reasons and because there are users that want to defend themselves from this kind of adversaries.

4   Application Example

In this section we will at first briefly discuss common approaches for attacks (further called *security issues*) providing for each of them the attacker class at least needed in order to efficiently execute the attack. Afterwards, we will give an overview on the strengths and weaknesses of existing or widely analyzed anonymizing networks. Furthermore, the maximum class of attacker that can be defeated by the corresponding technique will be provided according to the new attacker classification.

We clearly state that the following classification is done on the basis of our experience with anonymizing networks in theory as well as in practice. It is not to space out other possibilities to do the categorization in a completely different way. It is also well possible that extreme user behavior, future attacks or methods will change the level of protection. Thus, we expect a need to update the following lists in the future since they are done from today's perspective.

Furthermore, we distinguish three types of users depending on their behavior: *cautious*, *average*, and *unwary*. However, we intentionally do not describe these behaviors precise. Average behavior is achieved as it is understood in the common sense, e.g. through the usual web surfing. Under cautious users we understand those, that decide whether to use a specified service under concrete circumstances and send only a very limited number of messages. Unwary users do not care much about what are they doing. Further in this paper we will only consider *average* users. In general we expect cautious users to be able to protected themselves at least against attackers of one class ahead, while unwary users can be identified with much less effort.

Due to place restrictions we will not be able to explain all issues and techniques in detail. We thus rely on the reader to be familiar with the handled techniques and attacks, or follow up the referenced documents.

---

[6] The Mixmaster network is too distributed for attackers of these classes.

4.1   Security Issues

This section will provide a short overview on well known and analyzed security issues for anonymous communication systems. We gill give a short introduction and specify the class of attacker that is likely to draw significant advantage from the corresponding security issue. Note that most issues can be exploited in theory by an attacker with less power than given. But this typically relies upon fractional probabilities or pathological network structures.

We will use the following notation to describe the severeness of a single issue: after its main description we will add a number in brackets. The number denotes the class of attacker that is at least needed in order to *efficiently* mount this attack. By this we refer to the situation where an attacker of the concrete class succeeds in breaking the system (in order to de-anonymize a single average user) with some non-negligible probability.

It is an inherent property of the classification that several different attacks can be mounted by a single class of attacker. This is due to the fact that our work focuses on practical attacker representation, instead of fine-grained theoretical models that are needed to distinguish system properties of different techniques.

Denial of Service (0). A network should be as resistant as possible against (distributed) denial of service attacks and selfish nodes. The difficulty of this attack depends on the implementation characteristics of the service but can be as simple as attacking a couple of directory servers. If the anonymizing network is dis-functional due to a DoS-attack, some users switch to unprotected communication and thus give away the information they wanted to protect.

Hacking into a Node (0). This security issue deals with an active intrusion into the targeted node, possibly by means of security lacks in some services offered by the host. Having gained the access, the invader can overtake the control over the node (e.g. install spy software, etc.). This issue is of the great importance especially in anonymous communication systems because in most cases the majority of nodes is using the same software. Such a single vulnerability in this software can give an attacker the control over large parts of the network.

Analyze Application Layer Data (1). This attack analyzes any data that is transmitted from the client to the service provider without being changed, i.e. in the network layer above the anonymization layer. In most cases this refers to the data that is provided by the user through e.g. filling out a web form but can also include an analysis of HTTP- or email-headers that are transfered without modification. A good overview is given in [5].

Packet Counting and Delay Attacks (2). Packet Counting attacks work quite well on a small scale e.g. when the user is surfing the web [12]. However there are no studies that provide this analysis for current anonymity systems and it seems to be infeasible to apply this attack on other type of anonymizing networks like e.g. remailers. Additionally, packet counting can be thwarted by the use of dummy traffic.

On the other hand, delay attacks can be used to minimize the effect of dummy traffic and ease packet counting. In general, every attacker that is able to count the packets also has the possibility to delay them. However, this is not always true (e.g. in case of the shared medium). While delaying rises the chance for success, the attacker runs into the risk of being detectable.

End-to-End Traffic Analysis (3). Attackers that control a non-trivial part of the global network have a non-neglible probability of either controlling or observing a user's first node in the route and the exit point. Thus they are able to do end-to-end analysis.

$n-1$ **Attacks (4).** are also sometimes called Sybil attacks [8]. Depending on the system, it is not always necessary to deploy $n-1$ decoy nodes, it is rather sometimes sufficient to operate two nodes and wait until they happen to be introductory node, respectively exit point at the same time. In the Tor-network [7], this would suffice to break the system – of course, deploying more nodes raises the probability of the success. Thus, if an attacker of class 4 would like to do so, he would have the resources to run such an attack. Unfortunately, these attacks can only be thwarted by authentication schemes that are currently not solvable or deployable in practical systems.

Break Mixing (4). The same amount of influence on the network (i.e. observing the majority of nodes) is also needed to successfully mount a traffic analysis like described in [6,17][7].

Replay Attacks (5). In general, replay attacks are next to impossible to carry out against current implementations like e.g. ANON [3], Tor [7], and Mixmaster [19]. Thus, we grade the difficulty to the level where at least some cryptographic mechanisms have to be broken in order to replay messages. Since there are typically more efficient ways to learn the same information, we doubt that these kind of attacks can be seen in real systems.

4.2 Anonymizing Techniques

In this section we will consider the anonymity provided by several deployed anonymization techniques. We will specify the level of protection that is provided for an *average user* against known attacks. As one input we used the previous section 4.1 and weighted the classification according to the probability of success for each security issue with respect to a certain technique. But we also had to take implementation specific details into account as well as general weaknesses of the techniques.

In the following we will use a single number as notation to describe the maximum class of attackers that can be defeated by a certain technique.

Ants (2). The anonymizing networks Ants [2] and Mute [21] use ant-routing [11] to achieve anonymity. By their own judgement it can be broken under circumstances if the user is connected only to the nodes of the attacker. Additionally, there is no proof that the algorithms can't be degraded with

---

[7] See also section 4.2.

an attack similar to the one in [20]. The provided anonymity is at the level 2, whether it is also provided on the 3rd one it is not proven and therefore not known yet.

NDM, Onion Routing (3). NDM [9] and Onion Routing [7] can be defeated by end-to-end analysis, sybil attacks, packet counting attacks, and timing attacks [20]. While the risk of the first two can be thwarted and handled to a certain extend in the client's software or by cautious behavior, the latter two problems are more serious. On the other hand, it is still to be shown that the packet counting attacks can be successful in real networks with a high probability, and even if they are, they could be avoided with a software update (e.g. producing dummy traffic). Thus, we rate the protection of the average user to 3.

Mixing (3-4). Mixing can be added to Onion-routing in different flavors: fixed size batches, timed mixing, combinations of both [25], or stop-and-go mixes [15]. While the security gain by mixing is possibly questionable [6,17], it can still provide strong anonymity in open environments if users refrain from sending too much information in a single time interval [18].

We give no security level for Hash-Routing [27,10] and DC-nets [4] because there are no implementations that have a relevant user-base. Missing this, it is impossible to give a rating of their practical level of security.

5   Conclusion

There are currently no widely known implementation of anonymization network that would provide protection against arbitrary strong attackers. Thus, existing and commonly used attacker models, like e.g. global passive observer, are too strong in order to facilitate fine-grained analysis of todays practical systems. Such model is definitely needed for design and property evaluation of networks with strong anonymity properties. Researchers and end users, however, are also in need of a classification that allows differentiation for the methods that are used in today's implementations.

The proposed classification itself does not ease the risk analysis per se as it gives only the categories of attacker classes. The categorization of the difficulty of attacks or the protection provided by each single technique and its implementations is still subject to "manual" analysis. Hereby we mean, that it can only be used as a reference model to determine from which type of attacker the protection can be achieved. Even here it is possible that opinions differ and different people would classify in a different manner than we did.

We are aware that the classification has no analytical background, however it would be cumbersome and difficult to model real world entities. Additionally it seems currently computational infeasible to analytically proof the security provided by any implementation of theoretical techniques. Thus we had to rely on practical experience and not analytical arguments in favor of our criteria.

In this paper we proposed a classification of attacker types with regard to the attacker's influence on the network, the computational power and physical capabilities. It should not be seen as restriction since it is unlikely that the proposed entities and social stereotypes will be replaced or become irrelevant in the future, even if the underlying networks change. Furthermore, the provided classification can be easily communicated to the end-users and management level.

We hope that this document gives incentive to the community of researchers in the area of anonymous communication to think also about linking their theoretical models to realistic attackers and thus contributes to the discussion about measuring the quality of protection.

We'd also like to contribute with this work in future versions to classifications of attackers not only in anonymous communication systems but in the general field of IT-security.

References

1. Attacker Classification to Aid Targeting Critical Systems for Threat Modelling and Security Review. http://www.rockyh.net/papers/AttackerClassification.pdf, 2005. visited July 2006.
2. ANTS File Sharing. http://antsp2p.sourceforge.net/, 2005. visited Oct 2005.
3. O. Berthold, H. Federrath, and S. Köpsell. Web MIXes: A system for anonymous and unobservable Internet access. In H. Federrath, editor, *Proceedings of Designing Privacy Enhancing Technologies: Workshop on Design Issues in Anonymity and Unobservability*, pages 115–129. Springer-Verlag, LNCS 2009, July 2000.
4. D. L. Chaum. The Dining Cryptographers Problem: Unconditional Sender and Recipient Untraceability. *Journal of Cryptology*, (1):65 – 75, 1988.
5. R. Clayton, G. Danezis, and M. G. Kuhn. Real world patterns of failure in anonymity systems. In I. S. Moskowitz, editor, *Proceedings of Information Hiding Workshop (IH 2001)*, pages 230–244. Springer-Verlag, LNCS 2137, April 2001.
6. G. Danezis. Statistical disclosure attacks: Traffic confirmation in open environments. In Gritzalis, Vimercati, Samarati, and Katsikas, editors, *Proceedings of Security and Privacy in the Age of Uncertainty, (SEC2003)*, pages 421–426, Athens, May 2003. IFIP TC11, Kluwer.
7. R. Dingledine, N. Mathewson, and P. Syverson. Tor: The second-generation onion router. In *Proceedings of the 13th USENIX Security Symposium*, 2004.
8. J. Douceur. The Sybil Attack. In *Proceedings of the 1st International Peer To Peer Systems Workshop (IPTPS 2002)*, March 2002.
9. A. Fasbender, D. Kesdogan, and O. Kubitz. Analysis of security and privacy in mobile ip. In *Mobile IP, 4th International Conference on Telecommunication Systems Modeling and Analysis*. Nashville, March 1996.
10. S. Goel, M. Robson, M. Polte, and E. G. Sirer. Herbivore: A Scalable and Efficient Protocol for Anonymous Communication. Technical Report 2003-1890, Cornell University, Ithaca, NY, February 2003.
11. M. Günes and O. Spaniol. Ant-routing-algorithm for mobile multi-hop ad-hoc networks. In *Network control and engineering for Qos, security and mobility II, ISBN:1-4020-7616-9*, pages 120 – 138. Kluwer Academic Publishers, Norwell, MA, USA, 2003.
12. A. Hintz. Fingerprinting websites using traffic analysis. In R. Dingledine and P. Syverson, editors, *Proceedings of Privacy Enhancing Technologies workshop (PET 2002)*. Springer-Verlag, LNCS 2482, April 2002.

13. A. Hirt, M. J. Jacobson, and C. Williamson. Survey and analysis of anonymous communication schemes. Submitted to ACM Computing Surveys, Department of Computer Science, University of Calgary, December 2003.
14. J. D. Howard. *An Analysis Of Security Incidents On The Internet 1989-1995*. PhD thesis, Carnegie Mellon University, 1997.
15. D. Kesdogan, J. Egner, and R. Büschkes. Stop-and-Go-Mixes Providing Anonymity in an Open System. In D. Aucsmith, editor, *Information Hiding 98 - Second International Workshop*, pages 83 – 98. Springer Verlag, 1998.
16. D. Kesdogan and C. Palmer. The past present and future of network anonymity. Network Security, Special Issue of Computer Communications Journal, Elsevier, 2003.
17. D. Kesdogan and L. Pimenidis. The Hitting Set Attack on Anonymity Protocols. In *Proceedings of Information Hiding, 7th International Workshop*. Springer Verlag, 2004.
18. D. Kesdogan and L. Pimenidis. The Lower Bound of Attacks on Anonymity Systems – A Unicity Distance Approach. In *Proceedings of 1st Workshop on Quality of Protection, Colocated at ESORICS*, Milan, Italy, September 2005. LNCS.
19. U. Möller, L. Cottrell, P. Palfrader, and L. Sassaman. Mixmaster Protocol — Version 2. Draft, July 2003.
20. S. J. Murdoch and G. Danezis. Low-cost Traffic Analysis of Tor. Oakland, California, USA, May 2005. IEEE Symposium on Security and Privacy.
21. MUTE File Sharing. http://mute-net.sourceforge.net/, 2005. visited Oct 2005.
22. A. Pfitzmann. Security in IT Networks: Multilateral Security in Distributed and by Distributed Systems, October 2004. Script for the lectures "Security and Cryptography I+II".
23. A. Pfitzmann and M. Köhntopp. Anonymity, unobservability, and pseudonymity: A proposal for terminology. Draft, version 0.23, August 2005.
24. J.-F. Raymond. Traffic analysis: Protocols, attacks, design issues and open problems. In H. Federrath, editor, *Designing Privacy Enhancing Technologies: Proceedings of International Workshop on Design Issues in Anonymity and Unobservability*, volume 2009 of *LNCS*, pages 10–29. Springer-Verlag, 2001.
25. A. Serjantov, R. Dingledine, and P. Syverson. From a trickle to a flood: Active attacks on several mix types. In F. Petitcolas, editor, *Proceedings of Information Hiding Workshop (IH 2002)*. Springer-Verlag, LNCS 2578, October 2002.
26. A. Serjantov and P. Sewell. Passive attack analysis for connection-based anonymity systems. In *Proceedings of ESORICS 2003: European Symposium on Research in Computer Security (Gjøvik), LNCS 2808*, pages 116–131, Oct. 2003.
27. R. Sherwood, B. Bhattacharjee, and A. Srinivasan. P5: A protocol for scalable anonymous communication. In *Proceedings of the 2002 IEEE Symposium on Security and Privacy*, May 2002.
28. P. Syverson, G. Tsudik, M. Reed, and C. Landwehr. Towards an Analysis of Onion Routing Security. In H. Federrath, editor, *Proceedings of Designing Privacy Enhancing Technologies: Workshop on Design Issues in Anonymity and Unobservability*, pages 96–114. Springer-Verlag, LNCS 2009, July 2000.
29. M. Wright, M. Adler, B. N. Levine, and C. Shields. An analysis of the degradation of anonymous protocols. In *Proceedings of the Network and Distributed Security Symposium - NDSS '02*. IEEE, February 2002.

Author Index

Lecture Notes in Computer Science

For information about Vols. 1–4178

please contact your bookseller or Springer

Vol. 4221: L. Jiao, L. Wang, X. Gao, J. Liu, F. Wu (Eds.), Advances in Natural Computation, Part I. XLI, 992 pages. 2006.

Vol. 4219: D. Zamboni, C. Kruegel (Eds.), Recent Advances in Intrusion Detection. XII, 331 pages. 2006.

Vol. 4218: S. Graf, W. Zhang (Eds.), Automated Technology for Verification and Analysis. XIV, 540 pages. 2006.

Vol. 4217: P. Cuenca, L. Orozco-Barbosa (Eds.), Personal Wireless Communications. XV, 532 pages. 2006.

Vol. 4216: M.R. Berthold, R. Glen, I. Fischer (Eds.), Computational Life Sciences II. XIII, 269 pages. 2006. (Sublibrary LNBI).

Vol. 4215: D.W. Embley, A. Olivé, S. Ram (Eds.), Conceptual Modeling - ER 2006. XVI, 590 pages. 2006.

Vol. 4213: J. Fürnkranz, T. Scheffer, M. Spiliopoulou (Eds.), Knowledge Discovery in Databases: PKDD 2006. XXII, 660 pages. 2006. (Sublibrary LNAI).

Vol. 4212: J. Fürnkranz, T. Scheffer, M. Spiliopoulou (Eds.), Machine Learning: ECML 2006. XXIII, 851 pages. 2006. (Sublibrary LNAI).

Vol. 4211: P. Vogt, Y. Sugita, E. Tuci, C. Nehaniv (Eds.), Symbol Grounding and Beyond. VIII, 237 pages. 2006. (Sublibrary LNAI).

Vol. 4210: C. Priami (Ed.), Computational Methods in Systems Biology. X, 323 pages. 2006. (Sublibrary LNBI).

Vol. 4209: F. Crestani, P. Ferragina, M. Sanderson (Eds.), String Processing and Information Retrieval. XIV, 367 pages. 2006.

Vol. 4208: M. Gerndt, D. Kranzlmüller (Eds.), High Performance Computing and Communications. XXII, 938 pages. 2006.

Vol. 4207: Z. Ésik (Ed.), Computer Science Logic. XII, 627 pages. 2006.

Vol. 4206: P. Dourish, A. Friday (Eds.), UbiComp 2006: Ubiquitous Computing. XIX, 526 pages. 2006.

Vol. 4205: G. Bourque, N. El-Mabrouk (Eds.), Comparative Genomics. X, 231 pages. 2006. (Sublibrary LNBI).

Vol. 4204: F. Benhamou (Ed.), Principles and Practice of Constraint Programming - CP 2006. XVIII, 774 pages. 2006.

Vol. 4203: F. Esposito, Z.W. Raś, D. Malerba, G. Semeraro (Eds.), Foundations of Intelligent Systems. XVIII, 767 pages. 2006. (Sublibrary LNAI).

Vol. 4202: E. Asarin, P. Bouyer (Eds.), Formal Modeling and Analysis of Timed Systems. XI, 369 pages. 2006.

Vol. 4201: Y. Sakakibara, S. Kobayashi, K. Sato, T. Nishino, E. Tomita (Eds.), Grammatical Inference: Algorithms and Applications. XII, 359 pages. 2006. (Sublibrary LNAI).

Vol. 4200: I.F.C. Smith (Ed.), Intelligent Computing in Engineering and Architecture. XIII, 692 pages. 2006. (Sublibrary LNAI).

Vol. 4199: O. Nierstrasz, J. Whittle, D. Harel, G. Reggio (Eds.), Model Driven Engineering Languages and Systems. XVI, 798 pages. 2006.

Vol. 4198: O. Nasraoui, O. Zaiane, M. Spiliopoulou, B. Mobasher, B. Masand, P. Yu (Eds.), Advances in Web Minding and Web Usage Analysis. IX, 177 pages. 2006. (Sublibrary LNAI).

Vol. 4197: M. Raubal, H.J. Miller, A.U. Frank, M.F. Goodchild (Eds.), Geographic, Information Science. XIII, 419 pages. 2006.

Vol. 4196: K. Fischer, I.J. Timm, E. André, N. Zhong (Eds.), Multiagent System Technologies. X, 185 pages. 2006. (Sublibrary LNAI).

Vol. 4195: D. Gaiti, G. Pujolle, E. Al-Shaer, K. Calvert, S. Dobson, G. Leduc, O. Martikainen (Eds.), Autonomic Networking. IX, 316 pages. 2006.

Vol. 4194: V.G. Ganzha, E.W. Mayr, E.V. Vorozhtsov (Eds.), Computer Algebra in Scientific Computing. XI, 313 pages. 2006.

Vol. 4193: T.P. Runarsson, H.-G. Beyer, E. Burke, J.J. Merelo-Guervós, L.D. Whitley, X. Yao (Eds.), Parallel Problem Solving from Nature - PPSN IX. XIX, 1061 pages. 2006.

Vol. 4192: B. Mohr, J.L. Träff, J. Worringen, J. Dongarra (Eds.), Recent Advances in Parallel Virtual Machine and Message Passing Interface. XVI, 414 pages. 2006.

Vol. 4191: R. Larsen, M. Nielsen, J. Sporring (Eds.), Medical Image Computing and Computer-Assisted Intervention – MICCAI 2006, Part II. XXXVIII, 981 pages. 2006.

Vol. 4190: R. Larsen, M. Nielsen, J. Sporring (Eds.), Medical Image Computing and Computer-Assisted Intervention – MICCAI 2006, Part I. XXXVVIII, 949 pages. 2006.

Vol. 4189: D. Gollmann, J. Meier, A. Sabelfeld (Eds.), Computer Security – ESORICS 2006. XI, 548 pages. 2006.

Vol. 4188: P. Sojka, I. Kopeček, K. Pala (Eds.), Text, Speech and Dialogue. XV, 721 pages. 2006. (Sublibrary LNAI).

Vol. 4187: J.J. Alferes, J. Bailey, W. May, U. Schwertel (Eds.), Principles and Practice of Semantic Web Reasoning. XI, 277 pages. 2006.

Vol. 4186: C. Jesshope, C. Egan (Eds.), Advances in Computer Systems Architecture. XIV, 605 pages. 2006.

Vol. 4185: R. Mizoguchi, Z. Shi, F. Giunchiglia (Eds.), The Semantic Web – ASWC 2006. XX, 778 pages. 2006.

Vol. 4184: M. Bravetti, M. Núñez, G. Zavattaro (Eds.), Web Services and Formal Methods. X, 289 pages. 2006.

Vol. 4183: J. Euzenat, J. Domingue (Eds.), Artificial Intelligence: Methodology, Systems, and Applications. XIII, 291 pages. 2006. (Sublibrary LNAI).

Vol. 4182: H.T. Ng, M.-K. Leong, M.-Y. Kan, D. Ji (Eds.), Information Retrieval Technology. XVI, 684 pages. 2006.

Vol. 4180: M. Kohlhase, OMDoc – An Open Markup Format for Mathematical Documents [version 1.2]. XIX, 428 pages. 2006. (Sublibrary LNAI).

Vol. 4179: J. Blanc-Talon, W. Philips, D. Popescu, P. Scheunders (Eds.), Advanced Concepts for Intelligent Vision Systems. XXIV, 1224 pages. 2006.